THE M-1 HELMET
A History of the U.S. M-1 Helmet in World War II

THE M-1 HELMET

A History of the U.S. M-1 Helmet in World War II

MARK A. REYNOSA

Schiffer Military History
Atglen, PA

Acknowledgments

This book is the result of a great deal of assistance, by both individuals and agencies over a period of seven years. Most of those involved provided assistance or information that was often extremely hard to locate. Their help was greatly appreciated and without them this accurate history of the M-1 Helmet in World War II could not be told.

The following are those who deserve a special thanks and who made the effort possible: Chris Armold, Collector; Sidney J. Armistead, President, The Schuelter Company; Denise L. Bowler, Public Relations Manager, the BFGoodrich Company; Bridgestone/Firestone, Inc.; California State University, Northridge, Interlibrary Loan Office, Oviatt Library; Central Library, City of Los Angeles; Bernard M. Corona, Chief, Close Combat-Light & Heavy Division, Human Engineering Laboratory, U.S. Army Ordnance Center and School; Michel De Trez, Helmet Collector, Author, and Historian; Robert W. Fisch, Museum Curator, West Point Museum; Dallas W. Freeborn, Helmet Collector, Author, and Historian; Luther Hanson, Museum Specialist, U.S. Army Quartermaster Museum; Ed Hicks, Collector; Gary Howard, Collector, Author, and Historian; Keith R. Jamieson, M.D., Helmet Collector; Melissa A. N. Keiser, Chief Photo Archivist, National Air & Space Museum Archives Division; Super Image, Photo Support, Services, and Equipment; Clifford F. Lewis, Editor, Materials Engineering; Jon A. Maguire, Collector, Author, and Historian; Lisa A. Martellaro-Reynosa, Engineering Research and Support; Linda C. Martinez, Photo Support; Russell P. Nardi, Manager, Advertising and Marketing Materials, and Michael J. Cameron, Manager, Marketing Communications, McCord Gaskets, AE Clevite Inc.; Clayton R. Newell, Lieutenant Colonel, U.S. Army, Chief, Historical Services Division, Center of Military History; J. Pernell, Still Picture Branch, National Archives; George A. Petersen, Collector; Stuart W. Pyhrr, Curator, Arms and Armor, and the MMA Photo Library, Metropolitan Museum of Art; Edward J. Reese and DeAnne Blanton, Military Reference Branch, National Archives; Paula Reed, Librarian, Mine Safety Appliances Company; Crystal A.G. Reynosa, Photo Support; Michael R. Reynosa and Julie C. Reynosa, Mil-Spec Research, Engineering Research and Photo Equipment; Raymond G. Reynosa, World War II U.S. Army Veteran; Robert G. Reynosa and Olivia B. Reynosa, Research and Support; Charles A. Ruch, Historian, Westinghouse Electric Corporation; Kim Schroeder, General Motors Media Archivist, General Motors Corporation; John J. Slonaker, Chief, Historical Reference Branch, U.S. Army Military History Institute; K.L. Smith-Christmas, Material History Curator, United States Marine Corps Air-Ground Museum, Museum Branch Activities; St. Charles Public Library; Patrick A. Toensmeier, Editor, Modern Plastics Magazine; Anthony C. Wilsbacher, Captain, U.S. Army Reserve, Historian; Robert R. Yost, Director, Mohave Museum of History & Arts; and a special thanks to Bob Biondi and Peter Schiffer for publishing this book!

Dedicated to all those involved with the development and production of the M-1 helmet during World War II

Book Design by Robert Biondi.

Copyright © 1996 by Mark A. Reynosa.
Library of Congress Catalog Number: 96-67289.

All rights reserved. No part of this work may be reproduced or used in any forms or by any means – graphic, electronic or mechanical, including photocopying or information storage and retrieval systems – without written permission from the copyright holder.

Printed in China.
ISBN: 0-7643-0074-1

We are interested in hearing from authors with book ideas on related topics.

Published by Schiffer Publishing Ltd.
77 Lower Valley Road
Atglen, PA 19310
Please write for a free catalog.
This book may be purchased from the publisher.
Please include $2.95 postage.
Try your bookstore first.

```
623.441 R466m

Reynosa, Mark A., 1967-

The M-1 helmet
```

CONTENTS

Acknowledgments 4

Preface 7

Chapter 1   Background and M-1 Helmet Development 9

Chapter 2   Steel Helmet Body Production and Modification: 1941-1945 18

Chapter 3   Fiber Helmet Liner Production and Modification: 1941-1942 29

Chapter 4   Plastic Helmet Liner Development 34

Chapter 5   Plastic Helmet Liner Production and Modification: 1942-1945 36

Chapter 6   Procurement, Distribution, and the Complete Helmet 58

Chapter 7   Parachutist Helmet Production and Modification: 1942-1945 66

Chapter 8   Helmet Camouflage 76

Chapter 9   Field Modifications and Field Markings 90

Chapter 10   Toy and Work Helmets 107

Chapter 11   Helmet Body and Liner Production: 1951-1952 109

Bibliography 112

PREFACE

This book was written for the sole purpose of providing an accurate account of the M-1 Helmet in World War II. The scope of this book is to examine the development, production, modification, and procurement of the M-1 Helmet and its various components as produced during the period between 1941 and 1945. Previously little, if any, text has been written in this country about the M-1 Helmet. What has been written and published in the United States has amounted to no more than one or two pages, and often these accounts do not provide much information on the subject. Outside of this country, especially in England and France, much more has been written on the M-1 Helmet. While these foreign efforts were often sincere, the information published has often been speculative. My hope is that this book will provide many answers about the M-1 helmet, that have until this time, not been addressed in any publication. I have made every effort to obtain and use only primary sources in the compilation and writing of this text. The sources used were primarily United States Military sources, supplemented by a number of company write-ups from those major firms which participated directly in the development and manufacture of the M-1 Helmet, and articles from engineering industry trade magazines of the period.

The sample M-1 helmet body and helmet liner combination shown on the 13 August 1941. Note the rivet securing the sample helmet bodies web chin strap, and the additional snap fastener rivet cap on the sample fiber liner for the paratrooper web extension. (Courtesy of National Archives, 111-SC-122391)

CHAPTER ONE

BACKGROUND AND M-1 HELMET DEVELOPMENT

The history of the modern combat helmet has its origin in the First World War. Up to that time the steel combat helmet had been absent from the battlefield for many years. Its absence was due to the introduction of the musket and its ability to penetrate the armor of the day. In order to match the musket, the helmet had to increase its ballistic qualities. At the time the only way to do that was to increase the thickness of the helmet, and thus increase its weight. As a result the steel combat helmet began to weigh down the soldier and was finally dropped as a piece of combat equipment. It was not until World War I that the helmet made a comeback. This occurred during one particular incident in which a soldier had worn a metal bowl under his cap. When the soldier was struck by a rifle bullet in the head, the bowl protected the soldier from injury and possible death. This observation, when viewed against the heavy losses of troops from artillery fire, provided the inspiration to a French general to develop and issue protective metal headgear to his troops. Shortly thereafter, all armies were equipping their troops with metal helmets as essential combat gear.

During the First World War the United States forces had selected the British wash basin type helmet for use by their troops. The selection of the British helmet was made to expedite the manufacture and procurement of a standardized combat helmet for American troops. This helmet, when produced by American companies at the order of the United States Army, was designated the M-1917. The M-1917 helmet was not very comfortable to wear and did not provide adequate protection from shrapnel or other projectiles flung upward from the ground or approaching from the sides. The M-1917 also suffered from a lack of balance when worn by the soldier on the run. These short comings with the M-1917 helmet prompted the Army to continue to study other foreign helmets and to create experimental helmets of their own. Helmet study and experimentation continued after the World War and by 1920 the Army had an experimental helmet with the potential to surpass the M-1917 helmet. The new helmet was the first true "pot-shaped" helmet and it

Exterior view of the M-1917 helmet. (Courtesy of U.S. Army)

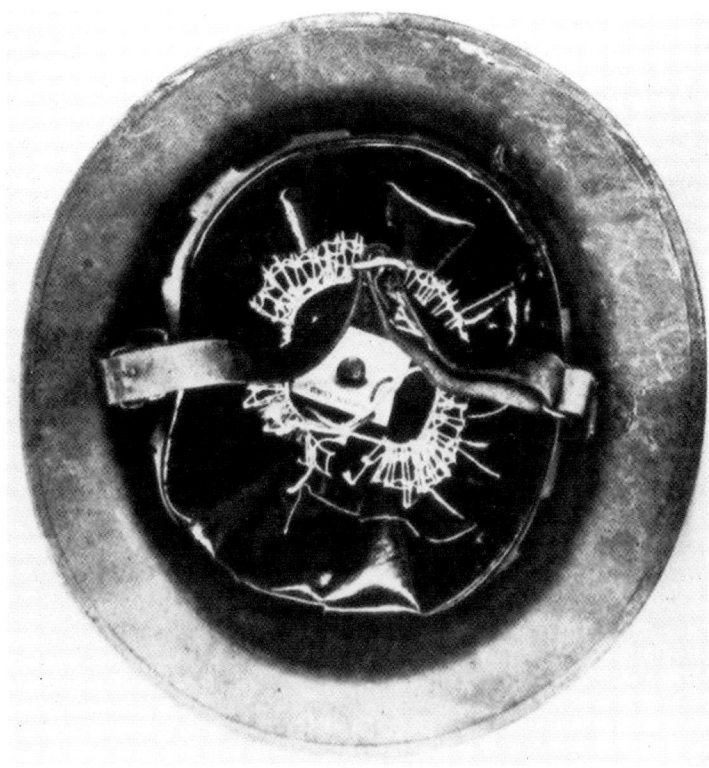

Interior view of the M-1917 helmet. (Courtesy of U.S. Army)

THE M-1 HELMET

Exterior view of the M-1917 helmet, ca. 1917-1918.

Interior view of the M-1917 helmet, ca. 1917-1918.

was designated the model No. 5A helmet. The model No. 5A helmet was just one of many experimental helmets developed near the end of the World War and it was designed by the armorers at the Metropolitan Museum of Art. Several service tests and preliminary trials were conducted at Fort Benning, Georgia in 1926 between the model No. 5A helmet and the M-1917 helmet. The tests indicated the M-1917 stood up to field conditions much better than the model No. 5A helmet. Specifically, the M-1917 was still ballistically superior in penetration, was not easily battered out of shape, was lighter in weight, and did not interfere with weapons firing. For these reasons, the Army continued to recommend the M-1917 helmet as the standard Army helmet. On 9 June 1932 the U.S. Army Ordnance Department decided to cease development, experimentation and testing of pot-shaped helmets. In January 1936, an order was issued to modify the M-1917 helmet by removing the old type of head pad and replacing it with new pads of an improved design. The improved head pads had been developed in 1934 in order to try to increase the comfort and stability of the M-1917 helmet. This newer version of the M-1917 helmet was designated the Helmet, M-1917A1.

U.S. ARMY NEED

In late 1940, the United States set out on a policy to have the best equipped army in the world. To meet this goal the United States Army was given the task of developing a new protective helmet for its soldiers. Knowledge gained by observing the events in the then active European war, suggested that the helmet which was used during the first world war would not provide the protection required. This was due to the fact that the current European war had proven to be a war in which projectiles could strike a soldier from any direction, and was not limited to the overhead bursting shell, which was common in the trench warfare of the first war for which the M-1917 was designed.

The job of developing this new helmet was given to the U.S. Army Infantry Board located at Fort Benning, Georgia. The Infantry Board immediately set out to establish a set of characteristics essential for the new helmet to meet. The Infantry Board stated the following in its report:

"Research indicates that the ideal shaped helmet is one with a dome-shaped top and generally following the contour of the head, allowing sufficient uniform head space for indentations, extending down in the front to cover the forehead with-

U.S. Army communications soldier wearing the M-1917A1 helmet during training, August 1941. (Courtesy of U.S. Army)

CHAPTER ONE: BACKGROUND AND M-1 HELMET DEVELOPMENT

Exterior front view of the M-1917A1 helmet. (Courtesy of George A. Petersen)

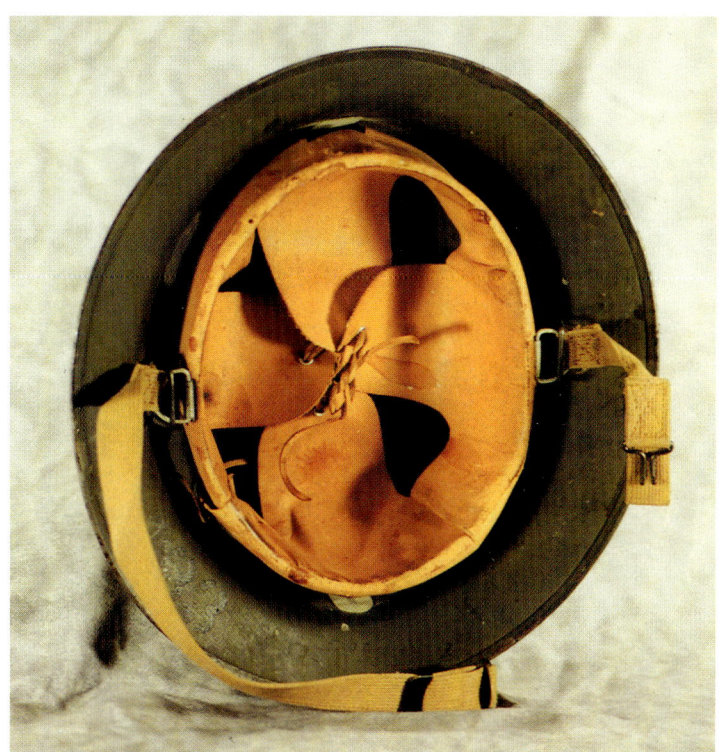

Interior view of the M-1917A1 helmet. (Courtesy of Chris Armold)

out impairing necessary vision, extending down on the sides as far as possible without interfering with the use of the rifle or other weapons, extending down the back of the head as far as possible without permitting the back of the neck to push the helmet forward on the head when the wearer assumes the prone position, to have the frontal plate flanged forward to form a cap-style visor, and to have the sides and rear slightly flanged outward to cause rain to clear the collar opening."

In addition to these requirements for more protection to the sides and the back of the head, the Infantry Board also established the following characteristics which were desired in the new helmet if it were to be standardized:

    a. The weight of the new helmet is not to exceed the weight of the M-1917A1 helmet, but it should be at least equal in ballistic qualities.

    b. The shape of the helmet is to be designed so as to stay on the head in skirmish runs and not to interfere with the firing of weapons.

    c. The helmet should provide maximum protection against shell and bomb fragments consistent with vision, hearing and weight.

    d. To improve comfort over the M-1917, the helmet should be padded so as to prevent headaches and discomfort to the wearer.

    e. The helmet should have two linings, one to protect the wearer from heat in the tropics and the other to protect the wearer from low temperatures in cold climates.

While the last characteristic did not materialize in the form of a standardized item, it did establish the fact that the Infantry Board was seeking to develop a new helmet with two distinct parts, a body and a liner.

The idea of a helmet and a separate liner had been suggested as early as 1932, but research into this goal was halted in that same year as the idea did not seem feasible at the time. With the need for a new helmet in late 1940, the Army again sought a two piece helmet. The reason behind the two piece helmet was that with such an item, the helmet liner could be worn by itself in forward areas in place of the standard garrison cap, and when combat was necessary, the soldier could quickly don the heavy steel body.

With this preliminary specification written out, development of the new helmet began in earnest.

STEEL BODY DEVELOPMENT

On 6 January 1941, shortly after the Infantry Board had issued its report, the U.S. Army Ordnance Department officially began to attack the problem of developing a new "pot-shaped" helmet. The Ordnance Department began to acquire an assortment of many different types of foreign helmets, then in use in the European war, for ballistic and metallurgical test purposes. At this time the Ordnance department requested the services of the personnel at the Metro-

THE M-1 HELMET

politan Museum of Art in New York. Help was sought from this museum because of their vast experience with ancient and modern helmets. Colonel Rene Studler, of the U.S. Army, was given the task of leading the effort at the museum. Among the staff who were influential in the design and development of the new helmet were Stephen Grancsay, who was then the curator of the Arms and Armor Department and a technical consultant to the Ordnance Department, and Leonard Heinrich, armorer at the Metropolitan Museum of Art.

After extensive research it was determined that the M-1917 helmet was the most suitable helmet to protect the top of the head. Since this was the case, it was decided to keep the M-1917 as the principle helmet with which to work in creating a new helmet. A standard M-1917 helmet was modified by first having the entire brim removed by trimming and then was extended along the sides and back by welding additional pieces of steel to the remaining dome shaped M-1917. A small visor was also welded to this rough model of the new helmet. Once this rough model was created at Fort Benning, a solid one piece experimental helmet was hammered out of Swedish Iron at the Metropolitan Museum of Art.

On 7 February 1941 the McCord Radiator and Manufacturing Company of Detroit, Michigan was awarded a contract by the Ordnance Department to produce sample dies and 200 sets of experimental sample helmet bodies and liners. McCord was awarded this contract since it was the only American company engaged in the production of combat helmets, having been given a contract by the Rock Island Arsenal to produce M-1917A1 helmets for the U.S. Army on 27 November 1940. Previously, McCord was involved in the manufacture of automotive gaskets, radiators, and cores. When McCord was given this second supplemental contract for sample dies, helmet bodies and liners, the Ordnance Department had also instructed McCord to stop production of the M-1917A1 by May 1941. McCord was requested to make this set of sample helmet bodies using Hadfield Manganese Steel. In order to manufacture the helmet bodies out of Hadfield Manganese Steel, McCord would have to develop presses capable of stamping out a helmet body in a single draw of 7". McCord met the challenge and the manufacture of the 200 sample helmet bodies produced by the sample dies was begun in March 1941.

LINER DEVELOPMENT

The development of the liner for the steel body was a joint project, of the Ordnance Department, the Quartermaster Corps, and private industry. The liner which was initially developed for the new pot-shaped helmet had its origins in the plastic football helmet and suspension invented and patented by John T. Riddell. Riddell had established a Chicago based company that manufactured a wide range of football equipment. The football helmet was first brought to the attention of the U.S. Army in early 1941 at Fort Benning, Georgia. At Fort Benning, preparations were under way for the training of parachute troops. A member of the Infantry Board, Colonel

U.S. Army parachutist wearing the Riddell Football helmet during parachutist training. (Courtesy of U.S. Army)

H. G. Sydenham was approached by Mr. Riddell, who suggested that his football helmet be used in the training of the parachute troops. Colonel Sydenham, along with others at the time, was responsible for developing a new combat helmet. As a result of this meeting, a modified form of the Riddell football helmet was used as the basis for the liner of the new helmet. The original version of this liner was created in Colonel Sydenham's kitchen by Riddell's son. The younger Riddell used a type of plastic called vinylite, which allowed him to mold the plastic liner into the experimental pot shaped helmet body by use of hot water to form the liner.

Soon after, the two pieces came together as the model for the new helmet assembly, the body and the liner. This new experimental helmet was then field tested at Fort Benning, and after receiving approval from the War Department, the Ordnance Department was allowed to continue with further experimentation. Since the Riddell experimental liner seemed to work, the company was asked to look into the possibility of manufacturing the liner out of plastic by an injection molding process. The company did so, but the resulting liner was too heavy to be considered at the time and until better techniques of plastic manufacture could be achieved, an alternative to plastic was sought. The suspension however was found to be superior in performance and was accepted. The suspension was

CHAPTER ONE: BACKGROUND AND M-1 HELMET DEVELOPMENT

Hawley tropical fiber hat, ca. 1943.

Interior view of the Hawley tropical fiber hat.

Close-up of the Hawley Products Company stamp found in the Hawley tropical fiber hat.

found to provide uniform space between the helmet and the head, and allowed head bands and neck bands of different sizes to be snapped onto the suspension in a manner which permitted the force of a blow on the helmet to be distributed equally to all parts of the head and neck bands.

By February 1941, a plastic alternative for the helmet liner was much sought after. McCord Radiator and Manufacturing Company, which was recently awarded a contract to produce 200 sets of experimental sample helmet bodies and liners, sought on its own to find a liner alternative. McCord, aware of the tropical helmets then being manufactured, turned to their manufacturer, the Hawley Products Company at St. Charles, Illinois. The Hawley Products Company had been producing their tropical helmets since 1933, when the company first manufactured and supplied the tropical helmets to concessionaires at the Chicago World's Fair. McCord, with no official orders or specifications asked Hawley Products to come up with a liner based on their tropical fiber helmet and incorporating the Riddell suspension system. The Hawley Products Company agreed to accept the task on the chance that it might later receive a sub-contract from McCord. The Hawley and McCord companies worked quickly and pooled their resources on the chance that what they were creating might just be what the Army wanted. In addition to Hawley and McCord, the Lilley Company and the George R. Carter Company, both of Detroit, Michigan, joined the development effort. These latter two companies were responsible for developing the suspension system based on the Riddell system, which had been recently licensed by the Army. In early April 1941, the companies completed manufacture of a set of 200 experimental sample liners which were immediately sent to the U.S. Army. By late April 1941, the companies received word from the U.S. Army Quartermaster that the helmet liner was acceptable. The gamble for the companies had paid off.

MODEL TS-3

This new combination of helmet body and liner was then designated as the Model TS-3. The TS-3 was designed to be manufactured in one size that would fit all head sizes. The helmet then underwent additional tests by the Infantry Board. The testing had shown that the TS-3 not only provided greater ballistic protection but that the extended areas at the sides and back also contributed to the TS-3 being more stable on the wearer's head. The TS-3 was also found not to interfere with the firing of weapons from any position, was comfortable to wear, and did not obscure the field of view to any great extent. Thus, the TS-3 had met most of the major goals set out by the Infantry Board in 1940. As a result of these tests, the Infantry Board found that the TS-3 was "most promising" and selected it for further ballistic and service tests at the Watertown arsenal.

The ballistics requirements for the TS-3 called for the helmet to meet and exceed those of the M-1917 helmet. The new helmet was to resist penetration of a 230-grain, caliber .45 bullet with a

THE M-1 HELMET

velocity of 800 feet per second. The helmet successfully met this requirement.

By late spring of 1941, the tests were over, and the final report was submitted by the Aberdeen Proving Ground. The conclusion of this report stated the following: "The experimental Helmet, TS-3 is ballistically superior to the requirements for a military helmet." Shortly thereafter, the standardization of the experimental Helmet, TS-3 as the Helmet, Steel, M-1, was approved on 9 June 1941.

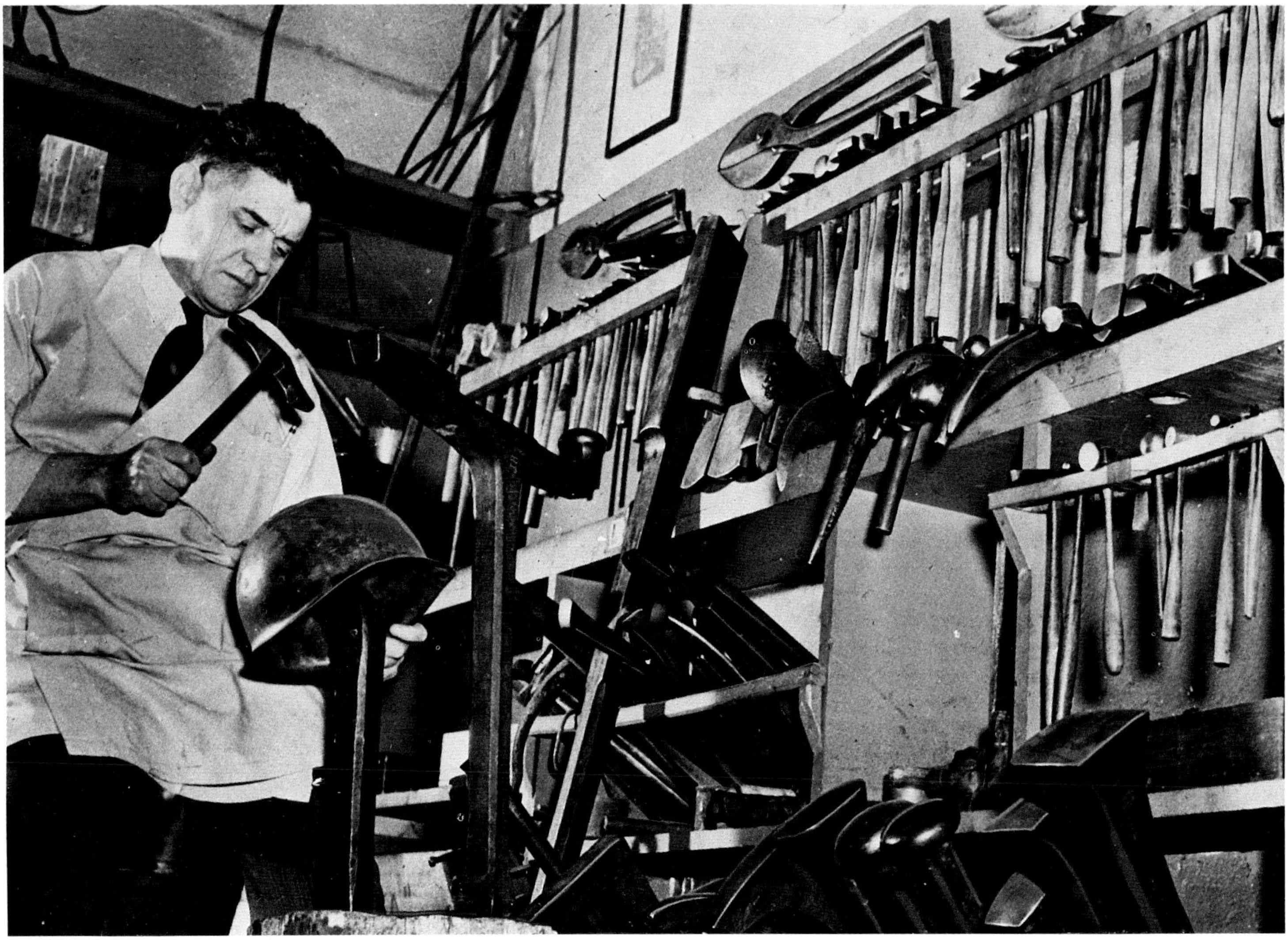

Leonard Heinrich hammered and shaped the experimental TS-3 helmet out of Swedish Iron at the Metropolitan Museum of Art Armorer's Workshop. (All Rights Reserved, The Metropolitan Museum of Art.)

McCord engineers examined the hand made model of the TS-3 helmet delivered to them by Army representatives. (Courtesy of McCord Gaskets, AE Clevite Inc.)

A McCord engineer closely examines one of the many laboratory samples. (Courtesy of McCord Gaskets, AE Clevite Inc.)

CHAPTER ONE: BACKGROUND AND M-1 HELMET DEVELOPMENT

Experimental TS-3 helmet body and fiber liner shown during Spring 1941. The TS-3 helmet would later be standardized as the M-1 helmet on 9 June 1941. Of special note are the rivets used to secure the helmet body's chin strap. This method of securing the chin strap to the helmet body was not used on the production helmet, instead the web chin strap would be sewn to the helmet body. Also note the fiber liner contains the original Riddell style webbing, which would be slightly changed on the production helmet, and the leather chin strap has two buckles instead of just one which would be used in production. (Courtesy of National Archives, 111-SC-120819)

M-1 helmet. Background, left to right, exterior view of the fiber liner; interior view of the fiber liner showing the assembled rayon suspension, rayon neck band, rayon and half leather head band, and leather chin strap; and the early model of the helmet body. Foreground, left to right, unassembled parts of the helmet including helmet body's cotton web chin straps, and helmet liner's rayon and half leather head band, rayon neck band, leather chin strap, and rayon suspension. (Courtesy of U.S. Army Ordnance Department)

THE M-1 HELMET

Front view of the M-1 helmet worn complete. (Courtesy of U.S. Army Ordnance Department via Keith R. Jamieson, M.D.)

Side view of the M-1 helmet worn complete. (Courtesy of U.S. Army Ordnance Department via Keith R. Jamieson, M.D.)

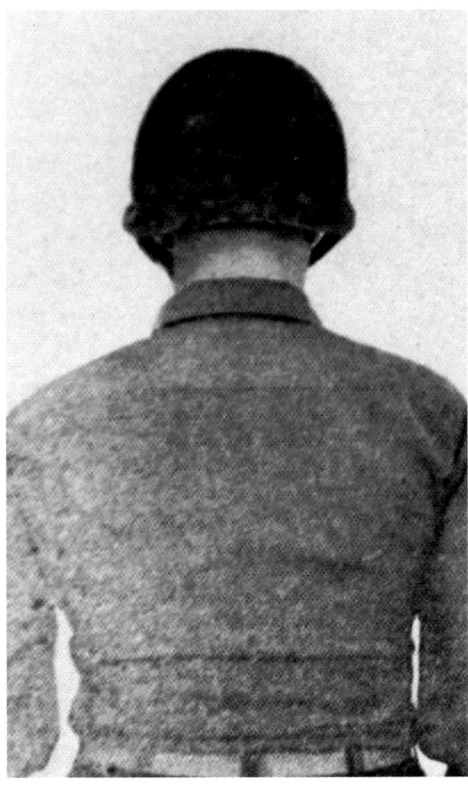

Rear view of the M-1 helmet worn complete. (Courtesy of U.S. Army Ordnance Department via Keith R. Jamieson, M.D.)

Side view of the M-1 helmet body and fiber liner being slip fitted together. (Courtesy of U.S. Army Ordnance Department via Keith R. Jamieson, M.D.)

Side view of the M-1 fiber liner worn alone as a field hat. (Courtesy of U.S. Army Ordnance Department via Keith R. Jamieson, M.D.)

Rear view of the M-1 fiber liner worn alone as a field hat. (Courtesy of U.S. Army Ordnance Department via Keith R. Jamieson, M.D.)

CHAPTER ONE: BACKGROUND AND M-1 HELMET DEVELOPMENT

Side view of the M-1 helmet body and fiber liner. (Courtesy of U.S. Army Signal Corps via George A. Petersen)

CHAPTER TWO

STEEL HELMET BODY PRODUCTION AND MODIFICATION: 1941-1945

Production of the steel body, which was an Ordnance item, began in the summer of 1941. The Ordnance Department had selected the McCord Radiator and Manufacturing Company of Detroit, Michigan, to produce the steel helmet bodies. McCord officially received its first M-1 helmet production contract on 26 June 1941. Earlier in the year McCord was producing the M-1917A1 and had halted production of that item in May, and immediately began to install new equipment necessary for the mass production of the M-1 steel body. The change in production came at a critical time when the Army was in need of helmets and this left the Army without any helmet production for a few weeks. The Ordnance Department picked up much of the criticism for the decision. When production of the helmet finally got under way there were still a number of problems in the process. Most of the initial problems dealt with the fact that the item had to be mass produced. The manufacture of the early set of 200 sample helmet bodies turned out to be a simple matter compared to the problems associated with mass production. With a mass produced item controlling tolerances, quality, and productivity proved to be very difficult. Eventually McCord solved many of the problems of mass producing the M-1 steel body, although it was only able to produce 323,510 helmet bodies by the end of 1941. By 1942, McCord had increased its production of the M-1 helmet bodies to about 5,000,000.

The helmet body produced by McCord, was manufactured from a single piece of Hadfield Manganese steel. Hadfield Manganese steel was an austenitic material which had the property of more than doubling its strength and hardness when cold worked. The Hadfield Manganese steel for the M-1 helmet body was produced

Manufacturing a section of the punch that would form the M-1 helmet body in a single draw. (Courtesy of McCord Gaskets, AE Clevite Inc.)

Grinding and polishing the die and punch used in forming the M-1 helmet body. (Courtesy of McCord Gaskets, AE Clevite Inc.)

CHAPTER TWO: STEEL HELMET BODY PRODUCTION: 1941-1945

Flat round sheets of Hadfield Manganese were placed under a press of high cobalt steel dies where they were formed by one draw into the M-1 helmet body, two at a time. (Courtesy of McCord Gaskets, AE Clevite Inc.)

Spot welding the stainless steel edge and chin strap loops to the M-1 helmet body. (Courtesy of McCord Gaskets, AE Clevite Inc.)

by the Carnegie-Illinois Steel Corporation and the Sharon Steel Corporation. The manganese steel ingot was produced by the standard open hearth method. The ingot was further reduced at these two plants into slabs and finally sheets. The final sheets were hand rolled to ensure proper thickness and quality and measured 34" wide by 70" long. The sheets were then packed rolled, heated up and quenched in cold water, austenitized. The individual sheets were then marked off with 16.5" circles and stamped with the austenite heat number. The sheet was then cut into two sheets with four circles apiece. Finally each blank or piece was cut out by circular shears, stamped with a shipment number, and packaged 400 to a crate. The heat number and shipment number were used to identify quality of steel and shipment lots throughout the manufacturing process.

The initial helmet body produced by McCord was manufactured from this single circular piece, or blank, of Hadfield Manganese, which was formed by a single draw and then trimmed. To this shell a thin piece of stainless steel was applied by spot welding to form a rim, with the rim butt occurring in the front of the helmet body. Also welded to the helmet were two stainless steel fixed loops that would hold the chin strap. The choice to use stainless steel was made in order to allow the entire helmet to be non-magnetic. This was necessary, so that the helmet would not cause an error in com-

The helmet body edge was beaded with strip stainless steel. (Courtesy of U.S. Army Ordnance Department via Keith R. Jamieson, M.D.)

Chin strap loops were spot welded to the helmet body. (Courtesy of U.S. Army Ordnance Department via Keith R. Jamieson, M.D.)

THE M-1 HELMET

Formed M-1 helmet bodies travel on a conveyor to the spray wash. (Courtesy of McCord Gaskets, AE Clevite Inc.)

Helmet bodies were spray painted in an automatic spray booth. (Courtesy of U.S. Army Ordnance Department via Keith R. Jamieson, M.D.)

pass readings during field operations. The stainless steel applied to the helmet was subjected to a number of tests at the Watertown arsenal, which insured that it did not detract from the M-1's ballistic qualities.

Once all the metal parts were applied, the helmet was placed in an acid wash to clean the helmet. After cleaning it was painted a shade of olive drab both inside and out. During the painting process the paint was mixed with cork and applied to the outer surface so that the helmet body would have a rough, non-reflective appearance. Finally, a two piece chin strap was sewn to the fixed loops. The khaki, olive drab shade no. 3, cotton web chin strap used, was initially developed in the latter part of the First World War for use with the M-1917. The strap consisted of a right and left part, the former being the longer of the two and having a buckle and a flat end clip attached. The brass flat end clip allowed adjustment of the chin strap's length. The buckle was a rectangular piece of brass that had two slits for the right chin strap and an opening with an inward pointing arrow that would attach with the brass chin strap release

Spraying the M-1 helmet body olive drab inside and out. The exterior paint was mixed with cork particles to produce a rough non-reflective appearance. (Courtesy of McCord Gaskets, AE Clevite Inc.)

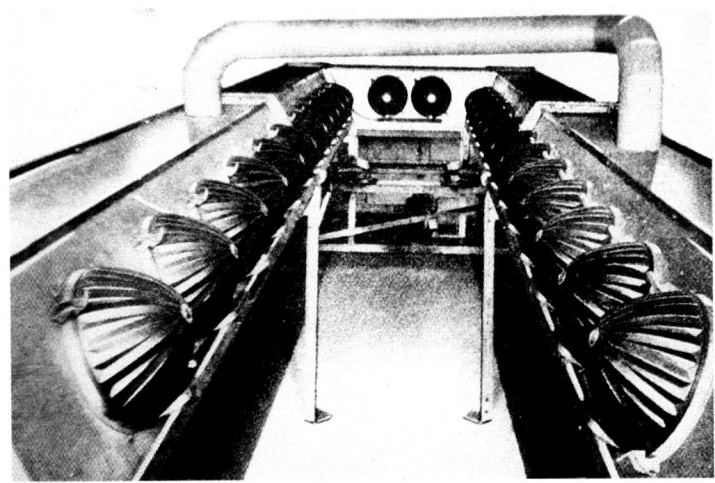

Drying the paint on the M-1 helmet body in an oven of infra-red lamps. (Courtesy of McCord Gaskets, AE Clevite Inc.)

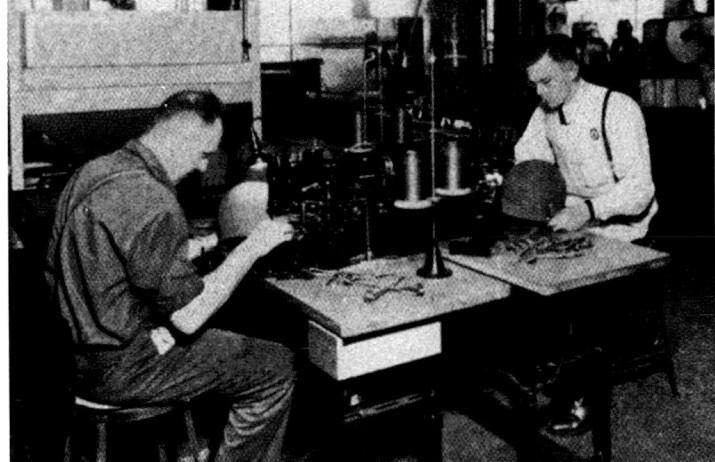

Cotton web chin straps were sewn to the helmet bodies after painting was complete. (Courtesy of U.S. Army Ordnance Department via Keith R. Jamieson, M.D.)

CHAPTER TWO: STEEL HELMET BODY PRODUCTION: 1941-1945

Exterior views of the M-1 helmet body, ca. July 1941-October 1943.

hook found on the left chin strap. Each of the brass components was given a rust and mildew inhibitor which produced a flat black coat like finish.

The completed M-1 helmet body, manufactured in only one size, weighed approximately 2.25 pounds and measured about 0.037" thick, 9.4" wide, 11.0" long and 6.9" deep.

The initial version of the M-1 helmet body suffered from a number of minor defects. One of the major problems was related to the manufacture of the initial helmets. The need to manufacture the M-1 body in a single 7" draw caused a serious problem. Initial drawing of the helmet had to proceed for some time before the right refinements could be made to the machines which manufactured the helmets. As a result, during the production, many of the bodies would simply break, usually in the area of the visor. Additional problems discovered through field use included the tendency of the helmet body to rust, and the development of "age cracks." The rusting was caused by a chemical reaction of both the paint and the cork when initially applied to the helmet body. A solution was found after much investigation which eliminated the problem by bonderizing the helmet before painting. The problem of age cracking was linked to poor quality steel and also a manufacturing oversight, which was remedied by annealing the helmet edge between the forming draw and the trimming operation during production.

Another problem was related to the design of the rigid non-flexible chin strap loops which tended to break off during repeated field use. The original loops did not give, and since they extended below the brim, they broke off easily when force was applied. The solution to this problem was to introduce a new flexible, hinged chin strap loop. The new loop was also constructed of stainless steel and was initially attached by two spot welds, and later three, to secure it to the helmet body. Tests were conducted on a number of helmets with the new loops and it was found that the new loops were fifty percent stronger than the original fixed loops. The tests also showed that no added discomfort was found when the new loops were used and that the previous method of sewing on the web chin strap was still acceptable. Production of the M-1 hel-

Top view M-1 steel body front edging butt, ca. July 1941-November 1944. This view shows the front butt on the stainless steel rim.

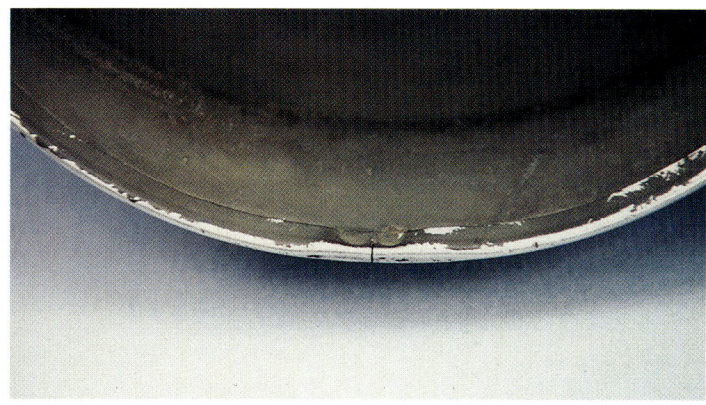

Interior view M-1 steel body front edging butt, ca. July 1941-November 1944. Note the spot welds holding the edging.

THE M-1 HELMET

M-1 steel body fixed chin strap loop, ca. July 1941- October 1943.

M-1 steel body chin strap buckle, ca. 1941-1943. This variation of the chin strap buckle was made of brass and had subtle raised features to hold the chin strap webbing.

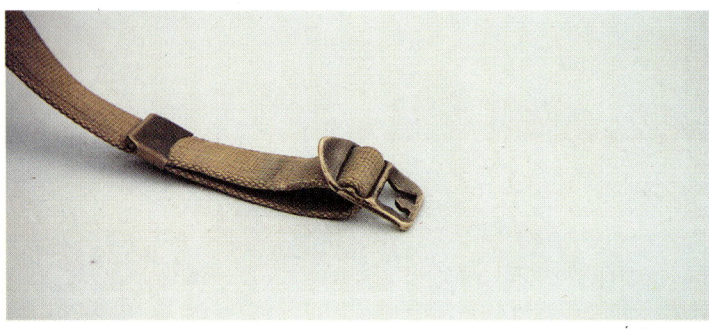

Side view of M-1 steel body chin strap buckle, ca. 1941-1943. This variation of the chin strap buckle was made of brass and had subtle raised features to hold the chin strap webbing.

M-1 steel body chin strap buckle, ca. 1941-1943. This variation of the chin strap buckle was made of brass and had prominent raised features to hold the chin strap webbing.

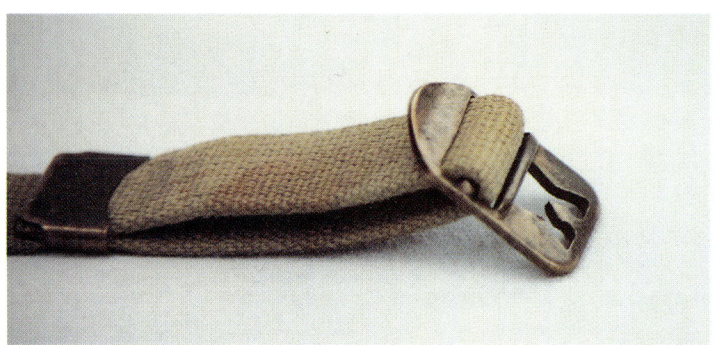

Side view of M-1 steel body chin strap buckle, ca. 1941-1943. This variation of the chin strap buckle was made of brass and had prominent raised features to hold the chin strap webbing.

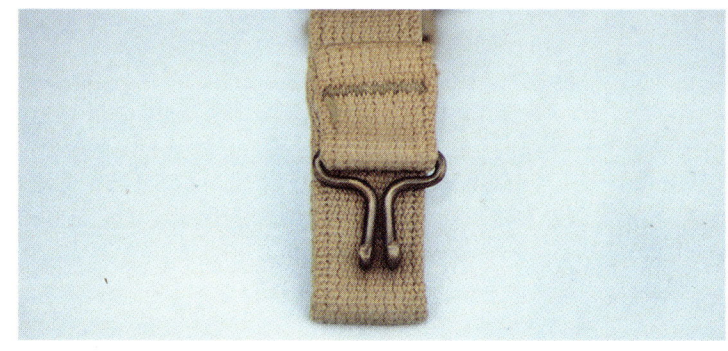

M-1 steel body chin strap release hook, ca. July 1941- October 1943.

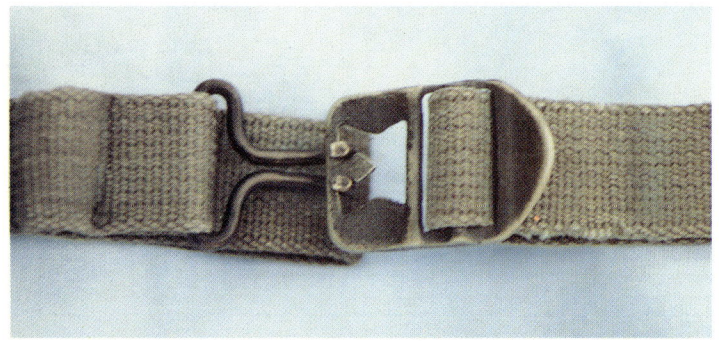

Fastened M-1 steel body chin strap buckle and release hook, ca. July 1941- October 1943.

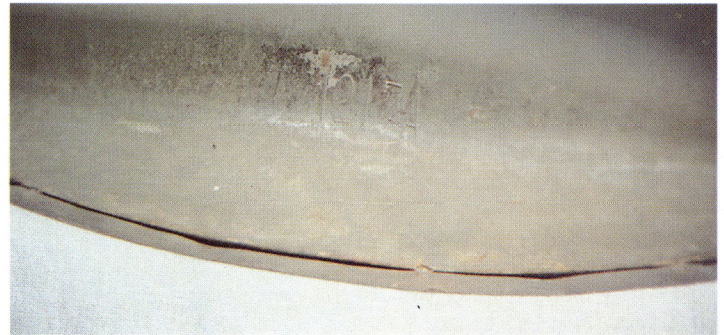

Close-up of stamp marking on the M-1 steel body. The stamp was placed under the visor of the helmet and represented the austenite heat number and the shipment number.

CHAPTER TWO: STEEL HELMET BODY PRODUCTION: 1941-1945

met bodies with the new hinged chin strap loop in place of the original design was begun around October 1943.

The original design of the M-1 helmet body called for the rim to be manufactured of stainless steel in order to give the helmet non-magnetic properties, but the stainless steel rim provided some serious problems as well. It was quickly found that paint did not adhere to the rim very well and as a result it tended to chip off, exposing the bright surface of the rim which reflected the sun and other light sources. The reflection caused a serious problem to the wearer as it might reveal the location of the wearer at a critical time in combat.

In May 1944, Army Service Forces requested that this problem be corrected as soon as possible. This resulted in an immediate recommendation for those helmets already in use. It stated that to correct the problem the stainless steel rim either be covered with Utility tape, pending development of a more suitable tape, or be sand blasted at a salvage repair depot to provide a rough surface which would be painted again. Soon after a better tape was developed to solve the reflectivity problem.

While this was a quick fix for helmets already in use, research was begun to provide a permanent solution in production. Initial efforts considered a number of chemical corrections which would allow the paint to better adhere to the rim, but it was finally decided to replace the stainless steel rim with another non-magnetic steel with the same non-reflective characteristics as the main helmet body. In October 1944, all production was converted to the Hadfield Manganese rim in place of the stainless steel rim. This production change of the helmet body permanently solved the problem of the light reflective rim. In addition to the conversion to Hadfield Manganese, it is probable that another production change or substitution was also enacted, which required the butt of the rim to be placed at the rear of the helmet body, as opposed to the front of the helmet body.

Procurement of the helmet bodies was principally the responsibility of the Ordnance Department, although the Quartermaster Corps did take over the duties for a short time in early 1943. Initially the Ordnance Department had selected a single contractor to produce the steel helmet body, the McCord Radiator and Manufac-

Exterior view of the M-1 helmet body, ca. October 1943-November 1944.

turing Company of Detroit, Michigan, but by summer of 1942, it was realized that a second company could greatly help the production effort and thus, the Schlueter Manufacturing Company of St. Louis, Missouri, was brought on as a second manufacturer. The Schlueter Manufacturing Company was awarded a contract for M-1 helmet bodies on 6 June 1942, although it did not manufacture a single helmet body until January 1943.

M-1 steel body flexible chin strap loop, ca. October 1943-August 1945.

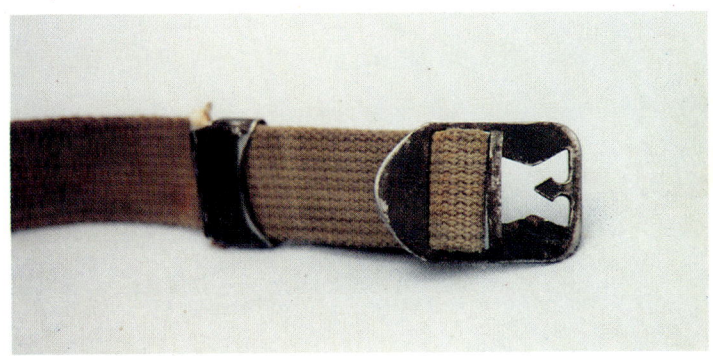

M-1 steel body chin strap buckle, ca. 1942-1943. This variation of the chin strap buckle was made of steel painted black and had prominent raised features to hold the chin strap webbing.

THE M-1 HELMET

M-1 helmet body. Late model with flexible chin strap loops. (Courtesy of U.S. Army Ordnance Department via National Archives, RG 156 Ordnance)

Production figures for the helmet bodies during World War II, were as follows:

Year	Quantity
1941	323,510
1942	5,001,384
1943	7,648,880
1944	5,703,520
1945	3,685,721

The total production of World War II M-1 helmet bodies reached 22,000,000 by VJ day, and ceased shortly after. McCord manufactured the bulk of these, approximately 20,000,000 helmet bodies, averaging 16,000 a day, while Schlueter produced approximately 2,000,000 helmet bodies, averaging 150,000 a month. It appears that McCord did not place any manufacturer's stamp mark on the M-1 helmet bodies they produced, however an S was stamped below the shipment number and heat number to identify M-1 helmet bodies produced by Schlueter.

M-1 helmet body chin strap buckle and release hook, ca. 1943. This variation was made of steel painted black.

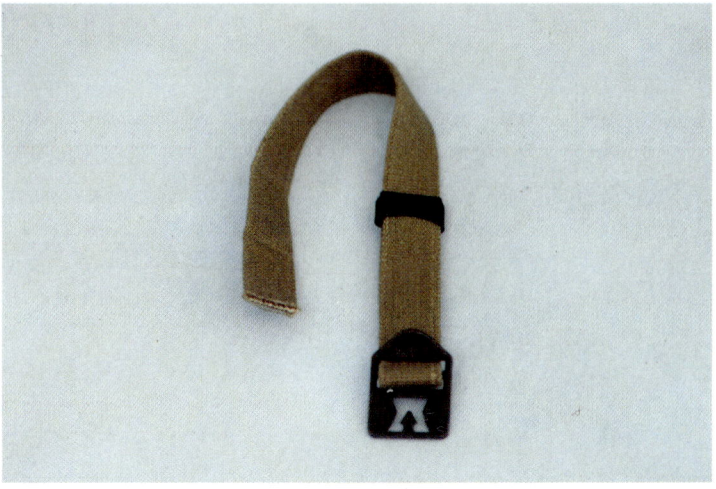

M-1 steel body cotton web chin strap, buckle and flat end clip.

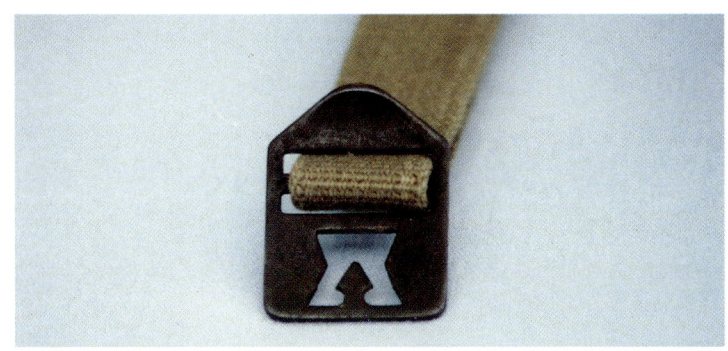

Close-up of the M-1 steel body chin strap buckle. This variation of the buckle is stamped brass coated with a mildew inhibitor, ca. 1944.

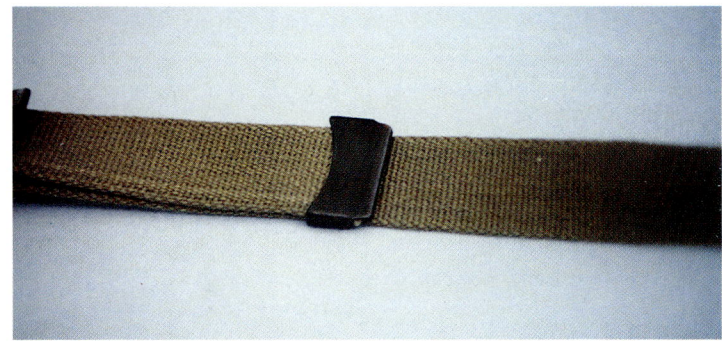

Front view close-up of the flat end clip.

Rear view close-up of the flat end clip.

24

CHAPTER TWO: STEEL HELMET BODY PRODUCTION: 1941-1945

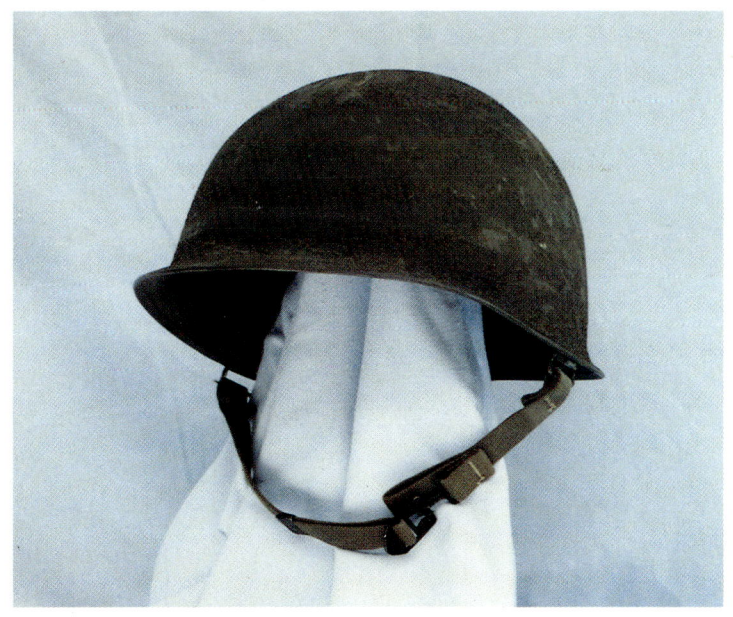

Exterior front left view of the M-1 steel helmet body, ca. November 1944-August 1945. This late variation steel helmet body has olive drab shade no. 7 webbing, brass fixtures, and a manganese rim with a rear edging butt. Right: Rear view of the M-1 helmet body, ca. November 1944-August 1945.

M-1 steel body rear edging butt, ca. November 1944-August 1945. This view shows the rear butt on the manganese rim. Also note the cork aggregate applied to the olive drab exterior paint.

During production, substitutions and other minor changes were allowed by the U.S. Army for the chin strap. The substitutions which occurred were the result of shortages of brass during the years of 1942 and 1943. In place of brass, the chin strap flat end clip, buckle and release hook were sometimes made of metal and given a coat of black paint. The only other change that occurred during the production of the helmet body's chin strap was the addition of a second webbing color, which was thought to have been olive drab shade no. 7, a dark green color. This change occurred in the fall of 1944.

ACCESSORIES AND COMPONENTS

As with any new item, problems were discovered in the field which required solutions in order to provide a helmet with more protection and easier maintenance. Among the items developed during World War II were a chin strap release accessory and a chin strap fastener component, which both later became standard production items on the M-1 helmet body in 1951.

The chin strap release accessory was developed in response to a report that came from the field in the spring of 1943. The report had stated that the blast of an exploding bomb or shell near the vicinity of a helmeted soldier would cause the M-1 helmet to capture the force of the blast, and in some instances this force was large enough to break the neck of the wearer. At the request of Army

M-1 helmet body with manganese rim and edging butt in the front, ca. November 1944-1945.

25

THE M-1 HELMET

 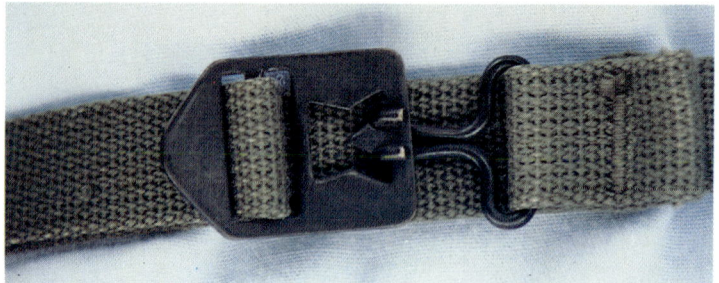

M-1 helmet body left chin strap with release hook, ca. November 1944-August 1945. Note that webbing on this example was olive drab shade no. 7 and the Bar Tack stitching is olive drab shade no. 3. Right: Fastened M-1 steel body chin strap buckle and release hook, ca. November 1944-August 1945.

Ground Forces, a research project was undertaken to develop a new chin strap device which would be released by the force of such a blast. Numerous tests were begun on a model of a human head constructed of wood and rubber and mounted on springs.

As a result of these extensive tests, which also included the use of slow motion pictures, a new chin strap release was designed and samples were made up for further testing. At the conclusion of these tests it was decided the chin strap release which would release at the force of fifteen pounds would be satisfactory. It was felt that this release would allow the helmet to come off when acted upon by a strong blast, but would also hold the helmet on during ordinary combat maneuvers.

This new chin strap release was soon manufactured and a mere thirty of these went overseas for approval and the rest were held pending acceptance. Army Ground Forces approved of the new release and requested several million more, although the item had not yet been standardized. The Release, Chin Strap, T1 was recommended to become a standard accessory for the M-1 Helmet in July 1944.

During the war, replacement parts for the helmet body became greatly needed, especially in July 1944, when it was reported that replacement of the helmet body chin strap had become acute. Due to shortages of field stitching equipment, it became necessary to turn in the helmet body when the cotton web chin strap was broken or frayed. A short term solution was found while a new removable chin strap was being developed. This short term solution recommended that field personnel could reattach their broken chin strap by use of any rivet, which did not have any reflective surface greater

The M-1 helmet body being used as a bowl for warming various liquids, including soups as shown by soldiers of the 148th Infantry Regiment on the island of New Georgia during August 1943. (Courtesy of U.S. Army Signal Corps)

CHAPTER TWO: STEEL HELMET BODY PRODUCTION: 1941-1945

T1 chin strap release for the M-1 helmet. (Courtesy of U.S. Army Ordnance Department via National Archives, RG 156 Ordnance)

M-1 helmet body chin strap buckle and release hook, with T1 chin strap release.

than 0.375". This recommendation permitted the soldiers to fix their own helmet without having to turn it in.

Development of a new removable chin strap lead to the design of a removable fastener, which was completed in September 1944, and sent to Army Ground Forces for further evaluation. The fastener was also suggested for use in the experimental T8 Helmet, which later became the Army Air Force's M-5 Helmet. In August 1945, the Fastener, Chin Strap, T1 was approved and adopted for use as a component for the M-1 Helmet.

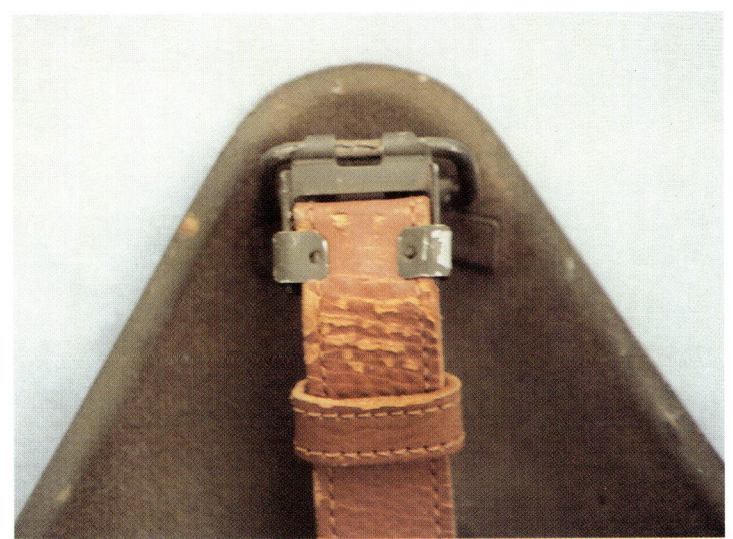

Close-up of the chin strap fastener as used on the M-5 helmet.

Close-up of the chin strap fastener as used on the M-5 helmet.

The U.S. Army Air Force M-5 helmet. Right: Interior view of the M-5 helmet.

27

THE M-1 HELMET

The M-1 helmet body was often used for washing, among other duties. (Courtesy of U.S. Army Signal Corps, 141396, via George A. Petersen)

CHAPTER THREE

FIBER HELMET LINER PRODUCTION AND MODIFICATION: 1941-1942

The first fiber liners were procured along with the first order of the M-1 helmet bodies, and as such the prime contractor was the McCord Radiator and Manufacturing Company. The U.S. Army Ordnance Department was initially responsible for the procurement of the fiber liners and the first contract was let out on 26 June 1941, as part of the first contract of M-1 Helmets. Soon after, McCord subcontracted the fiber liner body to its co-developer, the Hawley Products Company. While Hawley manufactured the fiber liner body, McCord supplied and installed the suspension to finish the liner at the McCord plant. Production at first was slow due to various problems, which required both McCord and Hawley to solve together. Once these problems had been solved at McCord's own expense, production of the fiber liner proceeded rapidly, as the fiber liner was extremely similar to the tropical helmet which the Hawley Products Company was already producing for the U.S. Army. In December 1941, responsibility for the procurement of the fiber helmet liner passed to the U.S. Army Quartermaster Corps.

The Army specifications for the fiber liner were written in October 1941, and stated that the liner was to be made of "two shells, each a one piece rigid fiber form, impregnated with varnish or other water insoluble and water repellent materials, securely cemented together with a suitable thermoplastic or thermosetting material which shall be insoluble in water." Once this body was formed, the fiber liner had a piece of olive drab gaberdine or twill smoothly cemented over it. The liner was designed and manufac-

Exterior side view of the M-1 fiber liner manufactured by the Hawley Products Company. Note the fraying of the exterior cloth on the front part of the rim, ca. late 1941-September 1942.

Exterior top left view of the M-1 fiber liner manufactured by the Hawley Products Company, ca. late 1941-September 1942. (Courtesy of Dallas W. Freeborn)

Exterior side view of an early Hawley Products M-1 fiber liner, ca. late 1941-September 1942. Exterior rivets of the same size indicate permanent leather liner chin strap. (Courtesy of Keith R. Jamieson, M.D.)

THE M-1 HELMET

Left to right, General Robert Richardson, Admiral Nimitz, and General Julian Smith on Tarawa, November 29, 1943. General Smith wears a Hawley M-1 fiber liner with the stenciled word "PAGE." (Courtesy of U.S. Marine Corps)

CHAPTER THREE: FIBER HELMET LINER PRODUCTION: 1941-1942

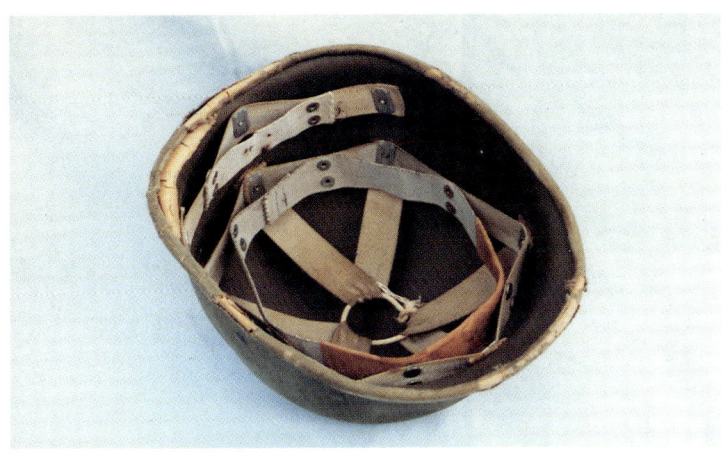

Interior front view of the Hawley M-1 fiber helmet liner, ca. late 1941-September 1942. Components consist of rayon Riddell suspension, steel perpendicular-sided washers, and permanent leather chin strap, not shown.

Close-up of the M-1 helmet liner perpendicular-sided metal washer, ca. late 1941-September 1942.

tured so that it would slip-fit into the M-1 steel helmet body. Three fittings were attached to this fiber liner body and included a suspension, neck strap, and an adjustable chin strap. The suspension and neck strap were both made of white rayon webbing, and secured to the liner by means of perpendicular-sided metal washers. Added to this liner were two inserts, which were issued at the time when the liner itself was issued to the wearer. These inserts were the head band and the neck band. The head band was produced in thirteen sizes, constructed of white rayon webbing and had a piece of leather sewn to the front half of it. The neck band was also constructed of white rayon webbing and was produced in three sizes. Both inserts were attached to the suspension and neck strap respectively, by means of snap fasteners. The leather chin strap was adjusted by a square two slit type buckle made of metal, and was

Interior view of the Hawley M-1 fiber helmet liner, ca. late 1941-September 1942. Components consist of rayon Riddell suspension, steel perpendicular-sided washers, and permanent leather chin strap. (Courtesy of Dallas W. Freeborn)

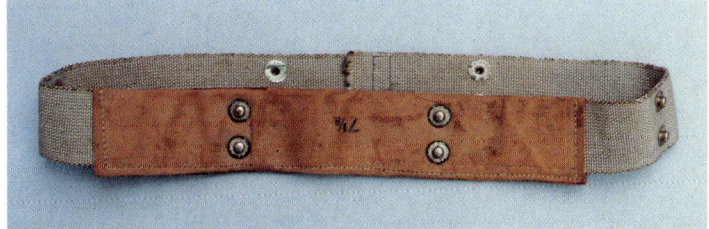

M-1 helmet liner sized head band. This sized head band is made of rayon webbing, leather and twelve snap fasteners, ca. 1942. The rayon sized head band came in thirteen sizes.

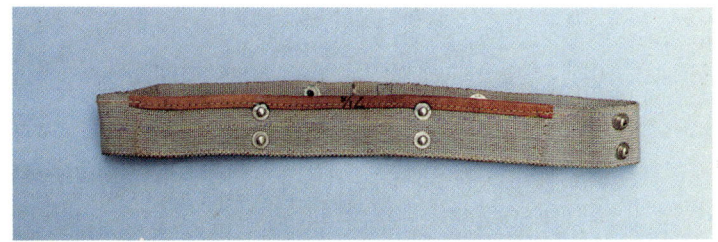

M-1 helmet liner sized head band. This variation of the sized head band was also made of rayon webbing, leather and twelve snap fasteners. Note that the leather on this variation does not wrap completely around the rayon webbing.

secured to the liner by permanent chin strap holders, which were riveted to the fiber liner body.

The fiber liner with the rayon webbing displayed very few markings. The only markings that did appear, were black numbers, indicating the size, which were stamped onto the head band and neck band inserts.

By June 1942, a new type of webbing had been developed to replace the white rayon webbing in the suspension system. This and other changes which are discussed in further detail in Chapter Five, resulted in approximately the last 672,000 fiber liners being produced with the new suspension system.

The Hawley Products Company manufactured 3,977,000 fiber liners for McCord. Of the total helmets, only three percent were sub-contracted to a second company, the General Fiber Company of St. Louis, Missouri. Production of the fiber liners did not begin until late 1941. The fiber liner continued in production until mid-November 1942, at which time production of the fiber liner was discontinued.

Bundle of rayon and leather snap fastened sized liner head bands. (Courtesy of Keith R. Jamieson, M.D.)

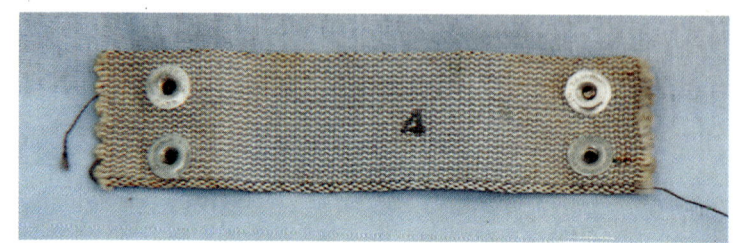

M-1 helmet liner neck band. This variation of the neck band was made of rayon webbing, ca. 1942. The character 4 represents the sizing number that was stamped on to these neck bands.

Left: Close-up of the M-1 helmet liner permanent chin strap holder, ca. 1942.

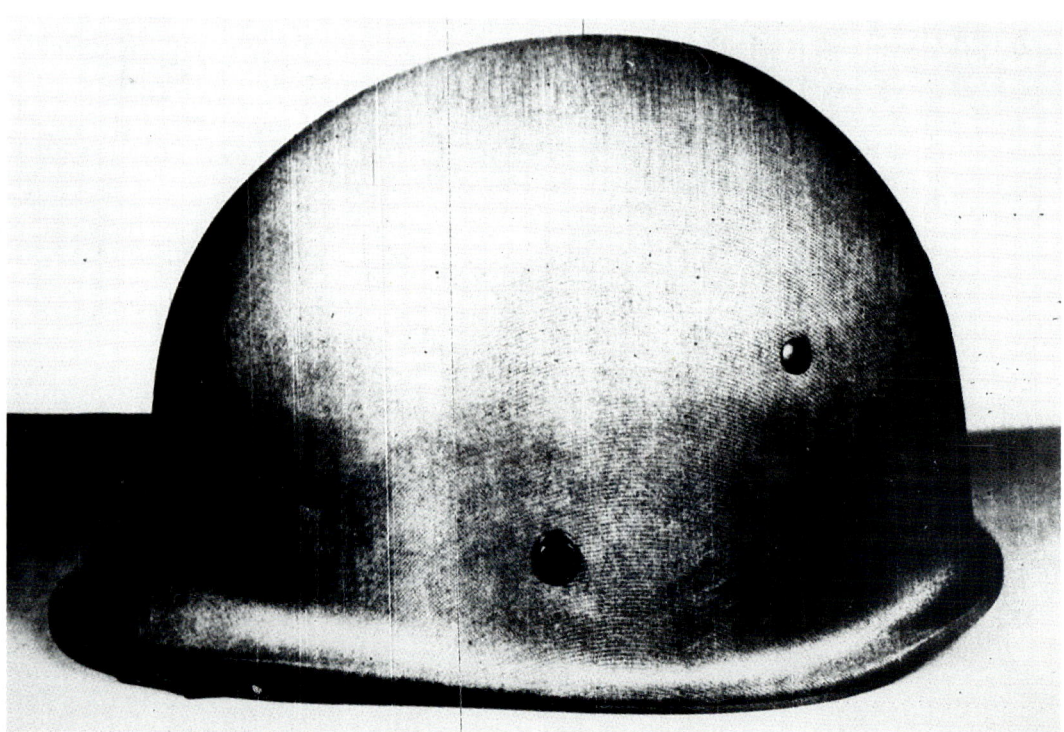

Right: Side view of the late version Hawley fiber liner. Note the larger rivet cap for the chin strap garter stud. (Courtesy of U.S. Army Quartermaster Corps)

CHAPTER THREE: FIBER HELMET LINER PRODUCTION: 1941-1942

Exterior top left view of a late version Hawley M-1 fiber helmet, ca. October 1942-November 1942. (Courtesy of Keith R. Jamieson, M.D.)

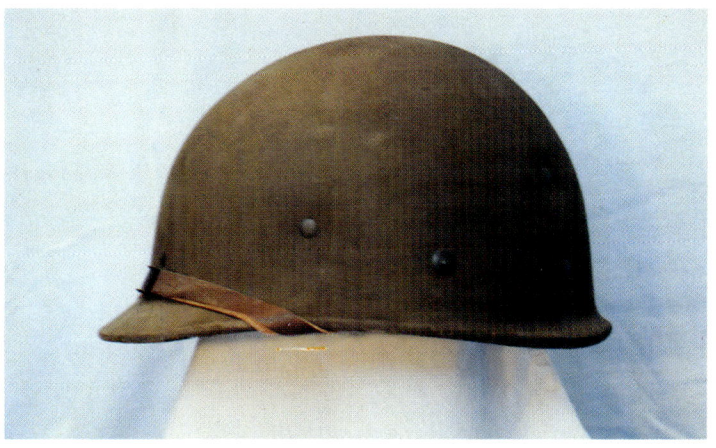

Exterior left side view of a late version Hawley M-1 fiber helmet, ca. October 1942-November 1942. (Courtesy of Keith R. Jamieson, M.D.)

Interior side view of a late version Hawley M-1 fiber helmet, ca. October 1942-November 1942. (Courtesy of Keith R. Jamieson, M.D.) Right: Interior front view of a late version Hawley M-1 fiber helmet, ca. October 1942-November 1942. Components consist of A washers, cotton olive drab shade no. 3 webbing for the suspension, neck band, adjustable head band, and a removable leather liner chin strap. (Courtesy of Keith R. Jamieson, M.D.)

Interior view of a late version Hawley M-1 fiber helmet, ca. October 1942-November 1942. (Courtesy of Keith R. Jamieson, M.D.)

CHAPTER FOUR

PLASTIC HELMET LINER DEVELOPMENT

The fiber liner was considered unsatisfactory almost from its initial production, but considering there were no other alternatives, it was the best possible liner available at the time if the M-1 was to become a standard issue item of the U.S. forces.

The fiber liner had been in production for just a few months and a number of problems in its performance began to surface. The liner most notably tended to absorb a great deal of water and thus lost most of its strength and durability. This would have grave consequences when the liner was issued to troops in the jungle, where the liner would literally become soggy and be of no use. Another problem which faced the liner was that it rather quickly lost its "neat" appearance due to the fact that the cloth outer layer was easily frayed and soiled. The Quartermaster Corps had tried to solve these and other problems, under the direction of Lieutenant Colonel Grice, but soon found that none of their attempts improved the characteristics of the fiber liner. With these problems continuing to plague the liner, the U.S. Army sought to continue research on the plastic helmet liner.

In July 1941, while the fiber liner was scheduled for production, research on the plastic helmet liner had begun in earnest. The responsible agency for the research was the Standardization Branch of the U.S. Army Quartermaster General. By August 1941, the Quartermaster sought to bring together industrial companies, which were familiar with the production of plastic items, to consider the possible replacement of the fiber liner by a laminated plastic liner. Also, samples of other types of civilian helmets were brought in for examination by the Army, and included miners' helmets, firemens' helmets and other industrial helmets. Companies present during this first meeting in August 1941 included the General Electric Company, Standard Products Company, Continental Diamond Fibre Company, St. Regis Paper Company, Hawley Products Company, the Inland Manufacturing Division of General Motors, the Mine Safety Appliances Company, and the Westinghouse Electric Company. The industrial companies participating in the development of the plastic liner were soon given a small contract and asked to produce sample plastic liners, which they felt would meet the current need. This small initial contract did not cover the cost for developing molding equipment, which caused Westinghouse to withdraw from the development effort. Those companies which pursued further experimentation, developed molding equipment at their own cost. During this time of development, the U.S. Army had not determined to any great extent the specification of the liner body and as such it was up to the individual companies to use their expertise in the field of plastics to come up with the best possible plastic liner body.

Among the first firms to submit plastic liners for evaluation were the Inland Manufacturing Division of General Motors, Dayton, Ohio, and the Mine Safety Appliances Company of Pittsburgh, Pennsylvania. Other firms submitting liners were the Gemloid Corporation, Elmhurst, New York; Standard Products Company, Detroit, Michigan; and Woodall Industries Inc., Detroit, Michigan. Hawley, aware of the plastic liner development, proposed its own new plastic liner, but a contract was never given to them to go forth with this idea.

The liners produced by these firms used some type of impregnating method applied to the interior reinforcing material. This method varied from company to company. Inland for its material used macerated canvas, crepe paper and fabric, pulp, and various wood barks to produce different samples. Mine Safety used canvas, while Gemloid, Standard Products, and Woodall used linen, felt and molded, not impregnated sisal fiber respectively. Mine Safety had vast helmet experience in the form of their industrial safety helmets, especially with a helmet which was similar to the future plastic liner and called the MSA Skullguard. The Skullguard helmet possessed properties very similar to what the Army desired for the plastic helmet liner. After experimental moldings were produced using a Quartermaster improved "Skullguard" formula, the Mine Safety Company agreed to release their helmet molding patents, royalty free, for the production of future Army plastic helmet liners.

In November 1941, the Office of the Quartermaster General Standardization Branch and the National Bureau of Standards had concluded that the best plastic helmet liner was one made of phenol

CHAPTER FOUR: PLASTIC HELMET LINER DEVELOPMENT

formaldehyde-impregnated cloth laminated by high-pressure. This conclusion was based on tests the Army conducted on the sample helmet liners submitted during the fall. The tests allowed the Army to establish the durability of each type of sample liner.

After 7 December 1941, at the insistence of the Quartermaster Corps, the Westinghouse Electric Company, Trafford, Pennsylvania, rejoined the development program for the plastic helmet liner. At this time the Quartermaster Corps was in a position to offer a development contract for both the product and the equipment for its production. Westinghouse assistance was sought for it too, like Mine Safety, had extensive experience in the production of safety and industrial helmets. Before the war, Westinghouse had more than twenty years of experience in resin impregnation and molding techniques and had been the principle contractor for the Micarta shell which was shipped to the Mine Safety Company for assembly and finishing. A major development had also occurred at Westinghouse between August 1941 and December 1941. Westinghouse engineers had developed new materials for the development of resin impregnated aircraft pulleys. These developments would be directly applicable to the plastic helmet liner.

Once Westinghouse was on board, the Office of the Quartermaster General requested that a lot of 600 sample liners be produced for further testing. All of these liners were to be produced on sample high-pressure molds developed by the Inland Manufacturing Division of General Motors, and a third of each were to be mixed according to the formula developed by Inland, Mine Safety, and Westinghouse to determine which of the three provided the best properties.

During this period of high-pressure molding experimentation, the LeGrand Daly Company, synthetic rubber and plastic engineers

Mine Safety Appliances Skullguard plastic helmet, ca. 1940.

working with the St. Clair Rubber Company, Marysville, Michigan, proposed a low-pressure method of molding a plastic helmet liner.

The results of the high-pressure molding tests were available by late January 1942, and at that time concern was still being raised over the need to produce a new helmet liner to replace the Hawley fiber liner which was already in production. The need for a plastic liner was clearly seen in the test report as it indicated that the plastic liners were far superior to the fiber liner. The report had also stated that the Westinghouse liner performed the best under service tests, especially ballistics, and was followed by the Inland liner.

On 3 February 1942, the Quartermaster Corps adopted the Westinghouse formula type, cloth impregnated helmet liner body. By 11 February 1942 a tentative specification had been written for those companies who would shortly receive the first contracts.

Interior view of the Mine Safety Appliances Skullguard plastic helmet. Right: Close-up of the Mine Safety Appliances stamp found in the Mine Safety Appliances Skullguard plastic helmet.

CHAPTER FIVE

PLASTIC HELMET LINER PRODUCTION AND MODIFICATION: 1942-1945

Contracts for the first plastic helmet liners were made by late February 1942. The initial production of the plastic helmet liner was contracted out to four companies and they were the Inland Manufacturing Division of General Motors, Dayton, Ohio; Micarta Division of the Westinghouse Electric and Manufacturing Company, Trafford, Pennsylvania; the Mine Safety Appliances Company, Pittsburgh, Pennsylvania; and the St. Clair Rubber Company, Marysville, Michigan. The St. Clair Company, had joined the development program late, and had also submitted their low-pressure helmet to the test trials. While their helmet liners did not perform as well as the Westinghouse and Inland liners, their method of low-pressure manufacture, instead of high-pressure, allowed for a greater number of plastic helmet liners to be produced more rapidly.

By April 1942, the second and last group of companies to receive plastic helmet liner contracts were chosen and they were the Capac Manufacturing Company, Capac, Michigan; Firestone Tire & Rubber Company, Akron, Ohio; Hood Rubber Company, Inc., Watertown, Massachusetts; International Molded Plastics, Inc., Cleveland, Ohio; and the Seaman Paper Company, Chicago, Illinois. All of these new manufacturers adopted the high-pressure style of production with the exception of the Hood Rubber Company, which adopted its own manufacturing process for the helmet, known as the ball-winding low-pressure method. Originally the Seaman Paper Company was awarded a contract to produce low-pressure helmet liners, but later discarded this method before production began, and switched to the high-pressure method with the help of Westinghouse.

Left: Dumbbell-Shaped segments of resin coated duck material were inspected at the St. Clair plant. (Courtesy of Modern Plastics Magazine, a division of The McGraw Hill Companies, Inc.) Right: The dumbbell segments were placed in the forming vessel, where they were preheated and cemented slightly together. (Courtesy of Modern Plastics Magazine, a division of The McGraw Hill Companies, Inc.)

CHAPTER FIVE: PLASTIC HELMET LINER PRODUCTION: 1942-1945

Left: The pre-formed St. Clair helmet liner was placed in the die cavity, and molded for ten minutes under 110 lbs. of steam pressure. (Courtesy of Modern Plastics Magazine, a division of The McGraw Hill Companies, Inc.) Right: The final operation at the St. Clair plant saw the helmet liner being trimmed and the edge sealed. The liner was then sent off to McCord for the installation of the suspension. (Courtesy of Modern Plastics Magazine, a division of The McGraw Hill Companies, Inc.)

During the summer of 1942, nine companies had received contracts to produce the plastic helmet liner. Seven of the companies would use the high-pressure method and the other two would use their own methods. The Army allowed the use of low-pressure due to the fact that liner production was badly needed and liners by this method were acceptable early on. These low-pressure liners were the first to be delivered in large numbers during the spring of 1942, while the first liners manufactured by the high-pressure method were not produced in large quantities until the fall of 1942. The ball-winding low-pressure method was allowed because it offered a less expensive form of production and did not use critical manufacturing resources which were needed elsewhere.

Left: Interior view of St. Clair low-pressure M-1 plastic helmet liner with Riddell suspension, permanent leather chin strap, perpendicular-sided washers and green interior finish, ca. April 1942. (Courtesy of Keith R. Jamieson, M.D.) Right: Close-up view of the permanent leather liner chin strap shown on an early version of the St. Clair M-1 plastic helmet liner. Visible is the square two slit type buckle made of metal and the permanent chin strap holder riveted to the liner. This type of chin strap was used on all the early liners. Also visible is the small recess cut made to the edge of the liner to provide clearance to the fixed chin strap loops found on the steel body. This feature was unique to the St. Clair liners. (Courtesy of Keith R. Jamieson, M.D.)

THE M-1 HELMET

Left: Exterior front right view of the St. Clair M-1 plastic helmet liner, May 1942-September 1942. This early plastic helmet liner has the dark olive paint job without texture. Right: Interior view of the low-pressure M-1 plastic helmet liner manufactured by St. Clair Rubber Company, ca. May 1942-September 1942. Note that the interior was no longer painted green.

LOW-PRESSURE ST. CLAIR PLASTIC HELMET LINER

The LeGrand Daly method employed by the St. Clair Rubber Company called for the plastic helmet liner to be formed under low-pressure using a steel mold to form the outer female cavity and a soft heat-resistant synthetic rubber and lignin inflatable bag to fit inside this cavity. This type of production overcame the time element involved in making female and male hardened steel dies for the high-pressure method and resulted in quicker production. The Army was eager to supplement production of the helmet liners until the high-pressure plastic liners could get into production, therefore it awarded St. Clair a contract in February 1942. The first deliveries of the St. Clair plastic helmet were received by the Army on the 22 April 1942. St. Clair had previously been involved in the manufacture of rubber floor mats for the automotive industry, as well as other rubber goods.

The impregnated material used in the LeGrand Daly method was a standard 8 ounce duck material pre-soaked in phenol-furfural resin. The material was more heavily coated on the side which would form the interior of the helmet liner. The material was then cut into dumbbell-shaped segments, six of which would form the helmet liner. Three of the segments were assembled to form the first layer of the pre-formed helmet in a forming vessel. The second layer of segments were then added to this first layer, each segment overlapping the other in the forming vessel. The entire pre-formed helmet required no stitching as the segments cemented to each other. Once pre-formed, the helmet was removed from the forming vessel and sent to the molding die cavities.

The pre-formed helmet liner was then placed in the steel mold die cavity. The steel mold cavity was jacketed to allow for steam circulation. The soft synthetic rubber inflatable bag was then placed in the pre-formed helmet, and the entire assembly was clamped down. The plastic helmet liner was finally formed by inflating the rubber bag and applying 110 pounds of steam pressure for ten minutes. The resulting helmet liner had a smooth outer surface and a rough inner surface as a result of the low-pressure applied by the rubber bag. The helmet liner was then trimmed and sealed at the

Left: Exterior front left view of the St. Clair M-1 plastic helmet liner, ca. September 1942. (Courtesy of Keith R. Jamieson, M.D.) Right: Exterior rear view of the St. Clair M-1 plastic helmet liner, ca. September 1942. (Courtesy of Keith R. Jamieson, M.D.)

CHAPTER FIVE: PLASTIC HELMET LINER PRODUCTION: 1942-1945

Interior views of the St. Clair M-1 plastic helmet liner, ca. September 1942. Components include cotton olive drab shade no. 3 suspension and neck band, unpainted steel A washers, and removable leather liner chin strap. This liner shows the transition from rayon to cotton as the cotton suspension was manufactured with snap fasteners to accept the rayon sized head band. (Courtesy of Keith R. Jamieson, M.D.)

edges. The helmet was then painted inside and out in dark olive drab and bake dried for six minutes. Finally the helmet was carefully inspected. After passing inspection the helmet was pierce punched and then the suspension, neck strap, insignia eyelet, and the studs for the chin strap were attached. The McCord Radiator and Manufacturing Company was subcontracted to supply and install the suspensions for the St. Clair plastic helmet liners.

Initially the St. Clair helmet liner was painted dark olive drab on its interior, later production changed and interior painting was completely dropped. Regardless of an interior paint job, the St. Clair plastic helmet liner always contained an interior crown stamp as required by the Army. The St. Clair stamp consisted of the yellow letters SC. The initial exterior paint job was also smooth and free of texturing, later models incorporated resin particles in the exterior paint to produce a textured effect.

If the helmet liners did not pass inspection, due to blemishes or other imperfections, they could either be discarded or trimmed and distributed to the civilian market as toy or work helmets. During the early months of low-pressure production, St. Clair was experiencing rejection rates of up to thirty percent. Imperfect, toy and work helmets will be discussed further in Chapter Ten.

Production of the St. Clair plastic helmet liners remained low with only 1,300,000 being produced before their contract was not renewed in early 1944.

Left: Close-up of the St. Clair Rubber Company stamp "SC." (Courtesy of Keith R. Jamieson, M.D.) Right: Interior view of the low-pressure M-1 plastic helmet liner manufactured by St. Clair Rubber Company, ca. October 1942. Components included the new cotton suspension in olive drab shade no. 3, unpainted steel A washers and the adjustable head band. Visible are the edges of the dumbbell shaped impregnated segments which make up the helmet. (Courtesy of Keith R. Jamieson, M.D.)

THE M-1 HELMET

Impregnated fabric being cut on a slitting machine for the Hood Rubber helmet liners. (Courtesy of BFGoodrich)

Ribbons of impregnated fabric were wound around a heated hollow oval mandrel to construct the pre-formed Hood Rubber helmet liners. (Courtesy of BFGoodrich)

Left: The wrapped Hood Rubber "water melon" shown before cross winding. After cross winding, the "water melon" is cut in half to form two pre-formed Hood Rubber helmet liners. (Courtesy of BFGoodrich) Center: A rubber bag acted as the inner mold plug. The pre-formed Hood Rubber helmet liner was placed over this bag. (Courtesy of BFGoodrich) Right: The pre-formed Hood Rubber helmet liner in place and ready for molding. (Courtesy of BFGoodrich)

HOOD RUBBER "BALL-WINDING" PLASTIC HELMET LINER

The Hood Rubber Company, a subsidiary of the BFGoodrich Company, entered the helmet liner picture as an experimenter. Prior to the war the company had experience in the manufacture of various rubber products, including rubber soles for shoes and inflatable rubber bags. The Hood Rubber Company was looking for a low cost alternative to produce plastic helmets for the U.S. Army and had received a contract to do so on 18 April 1942. The low cost alternative, it was hoped, would use inexpensive equipment and make use of minimal critical materials and steel. After eight months of work, the company had its new low cost method, the "Ball-Winding" process.

The material used in the Hood "Ball-Winding" process was 4 ounce high count cotton sheeting impregnated with resin. The material was then cut into long ribbons, both 2" wide and 0.75" wide, which were wound onto spools. The ribbons of material were then wound around a heated mandrel shaped like a watermelon. The mandrel was constructed of two hollow metal halves. Before wrapping was started, the mandrel was covered with talcum powder, to

CHAPTER FIVE: PLASTIC HELMET LINER PRODUCTION: 1942-1945

Left: The finished Hood Rubber helmet liner being removed from the mold. (Courtesy of BFGoodrich) Center: Trimming operation on the molded Hood Rubber helmet liner. (Courtesy of BFGoodrich) Right: The Hood Rubber helmet liner had 12 holes punched by two machines to prepare the liner for its interior fittings. (Courtesy of BFGoodrich)

Left: Suspension harnesses being riveted to the Hood Rubber helmet liner. (Courtesy of BFGoodrich) Right: View of the interior of the Hood Rubber helmet liner. The suspension is shown in place. The liner chin strap was added after the exterior of the liner was spray painted. (Courtesy of BFGoodrich) Below: A conveyor belt moved the Hood Rubber helmet liners through a spray booth where the liners were painted olive drab. The conveyor then took the painted liners through a tunnel of infra-red lamps where the paint was dried. (Courtesy of BFGoodrich)

THE M-1 HELMET

prevent sticking, and heated to 160 degrees Fahrenheit. The 2" ribbons were wound first, and then the 0.75" ribbons were cross wound. The ribbon covered "melon" was then cut to form two pre-formed helmet liners. Next, ducking material was attached around the edge to form a brim collar, and finally a 3.5" circular segment of impregnated material was placed in the interior crown.

The pre-formed helmet liner was then placed in a rubber inflatable bag to prepare it for molding. Above the rubber bag and pre-formed helmet was the steel female mold frame. The mold frame was lowered and the rubber bag inflated with hot air injected at 250 p.s.i., to cure the assembly. After six minutes the plastic helmet liner was cured. After molding was complete, the helmet liner was trimmed on a three way cutter. The liner edges were then burnished and sealed. The liner then placed on a punching press which punched two sets of six holes to admit the suspension, neck strap, insignia eyelet, and the studs for the chin strap. These hardware items were then riveted to the liner. Finally the liner was sprayed with olive drab paint containing texturing particles, and dried under infra-red lamps.

The Hood Rubber plastic helmet liner contained an interior crown stamp as required by the Army. The Hood Rubber stamp consisted of the silver letters HR.

Production by the Hood Rubber Company also remained low with only 206,000 plastic helmet liners being manufactured before their contract was not renewed in early 1944.

HIGH-PRESSURE PLASTIC HELMET LINER

The high-pressure plastic helmet was produced by the following firms: Inland Manufacturing Division; Westinghouse Electric and Manufacturing Company; the Mine Safety Appliances Company; the Capac Manufacturing Company; Firestone Tire & Rubber Company; International Molded Plastics, Inc.; and the Seaman Paper Company. The high-pressure manufacturing method was the method most desired by the Army as it produced a helmet which met and exceeded their requirements and specifications for a helmet liner. This type of production required the manufacture of female and male hardened steel dies for curing the plastic helmet liner via high-pressure applied by steel. Although initial contracts for the high-pressure plastic helmet liner were issued in early February 1942, deliveries of the first production high-pressure plastic helmet liners did not begin until May 1942. Even then, production of the high-pressure plastic helmet liner remained at a trickle until manufacturing problems could be worked out and mass production begun in the fall of 1942. Westinghouse was the first manufacturer to deliver high-pressure M-1 plastic helmet liners in May 1942. Through the summer of 1942, Westinghouse was followed in first deliveries by the Inland Manufacturing Division of General Motors, Firestone, Seaman Paper Company, and Mine Safety Appliances, respectfully. Capac Manufacturing and International Molded Plastics were not able to make their first deliveries until after 30 September 1942.

Interior view of the Hood Rubber Company low-pressure helmet liner, ca. early 1943. Components included the cotton suspension in olive drab shade no. 3, unpainted steel A washers, adjustable head band and removable chin strap. (Courtesy of Dallas W. Freeborn)

Close-up of the Hood Rubber Company stamp "HR." (Courtesy of Dallas W. Freeborn)

CHAPTER FIVE: PLASTIC HELMET LINER PRODUCTION: 1942-1945

Left: Die-cut impregnated duck sections were stapled to form a hat-shape for the plastic helmet liner. (Courtesy of U.S. Army Quartermaster Corps) Center: Three of the hat-shapes were placed, one inside the other, inside a liner mold prior to being formed. (Courtesy of U.S. Army Quartermaster Corps) Right: Conveyors carried hat-shaped pre-forms and newly molded helmet liners to and from the molding press. (Courtesy of U.S. Army Quartermaster Corps)

The plastic helmet liner, when produced using the high-pressure method, was manufactured by the following method. Cotton ducking material was first impregnated with resin and then cut into wedge, leaf, or reversed pie shaped segments. The individual companies were free to use their own inventiveness in preparing the pre-form helmet. These shaped segments were then arranged and stapled together to create a pre-formed helmet liner. An extra piece of impregnated duck material was added to the interior crown for strength. The pre-formed helmet was then placed into a steel mold. In the steel mold, high-pressure force was applied and the helmet was then allowed to cure. The pre-form was subjected to two minutes of 150 tons of pressure at 220 degrees Fahrenheit. The result was a hard, shaped helmet body. This hard, shaped helmet body had a smooth finish both inside and out. The next operation called for the removal of excess flash from around the helmet liner edge, this was pre-formed by a punch press. Next the edges of the helmet were sealed by burnishing. This helmet body was then finished by first being pierced by a hole punch, the holes being needed for the attachment by riveting of the suspension, neck strap, insignia eyelet, and the studs for the chin strap, and then secondly by the riveting of these items. The final process called for the helmet to be painted and baked dried. The paint was only applied to the exterior of the liner and was olive drab in color. In some plants, painting was accomplished using automatic paint spray machines, while other plants used hand paint spray. Baking of the paint was performed by either using an oven of infra-red lamps for two minutes, or using a horizontal drying ovens for fifteen minutes. The infra-red lamps used in this final operation, had recently been developed at the Bloomfield Lamp Division of Westinghouse.

The finished liner varied in thickness but averaged about 0.082" thick, and measured 8.6" wide by 10.6" long. The weight of the liner with all of its components assembled was 0.75 pounds. The helmet liner shell, when finished, also contained a small molded marking on the interior of the crown to indicate the manufacturing

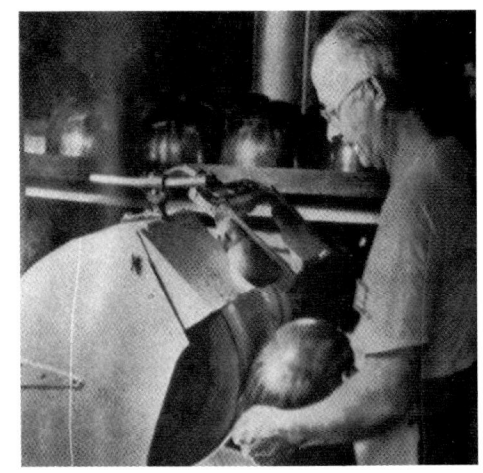

Left: The pre-forms were placed in a press and mold cavity for the high-pressure molding of the plastic helmet liners. (Courtesy of Westinghouse Electric Corporation) Center: A punch press removed all excess flash from the helmet liner in one operation. (Courtesy of U.S. Army Quartermaster Corps) Right: The edge of the helmet liner was smoothed during a burnishing operation. (Courtesy of U.S. Army Quartermaster Corps)

THE M-1 HELMET

Here, the air cylinder toggle piercing press punched 12 holes into the liner to admit the various rivets and eyelet at the Westinghouse Micarta Division during October 1943. (Courtesy of Westinghouse Electric Corporation)

company. The interior, which was not painted, possessed a brown striped design which was attributed to the impregnated duck fabric segments used in the high-pressure manufacturing method.

PLASTIC HELMET LINER COMPONENTS AND MODIFICATIONS

The plastic helmet liner initially produced incorporated the Riddell suspension system with the white rayon webbing as used in the earlier Hawley fiber liner. Approximately 3,600,000 plastic helmet liners were manufactured in this way before production was changed to the new cotton suspension system in June 1942. Development of a new suspension system had begun in January 1942, and a new system was sought because the Riddell system was discovered to have a number of unsatisfactory characteristics. The white rayon webbing used in the suspension tended to stretch, and thus destroyed the fit of the helmet. The perpendicular sided washers used to secure the suspension to the helmet were also troublesome as they often hurt the wearer's head. The biggest problem by far was the need for thirteen sizes for the head band. Not only did the Army not like this requirement, as it was difficult to procure just the right amount of the sizes needed, but the means of fastening the head band through use of six sets of two snap fasteners, also provided discomfort to the wearer.

Solutions to these various problems were worked out early in 1942, through the cooperation of the U.S. Army Research and Development Branch of the Military Planning Division and private industry. The rayon webbing was to be replaced by a cotton webbing of khaki color, olive drab shade no. 3, which proved more durable, and was manufactured in a single Herringbone twill weave. The single Herringbone twill weave produced a v-type pattern in the webbing. To secure this new webbing to the liner a new washer was recommended and developed by the Westinghouse Electric Company. This new A-shaped washer replaced the older perpendicular-sided washers. The hammock style suspension still retained a shoelace for adjustment as it did when the suspension was made of rayon. The basic Riddell web-type suspension remained but many of its smaller components were changed.

A new adjustable head band, needed to eliminate the thirteen sizes previously used was developed by the Scholl Manufacturing Company. The head band was made adjustable by use of a two piece bar buckle. The new head band was also designed to have vegetable tanned calfskin sewn on one side for its entire length, except at its ends, to allow adjustment by the bar buckle. Mr. Garrett C. Skinner, manager of Scholl, had developed a new chin strap buckle for the liner and a new spring clip used to attach the head band to the suspension. The new buckle replaced the older square

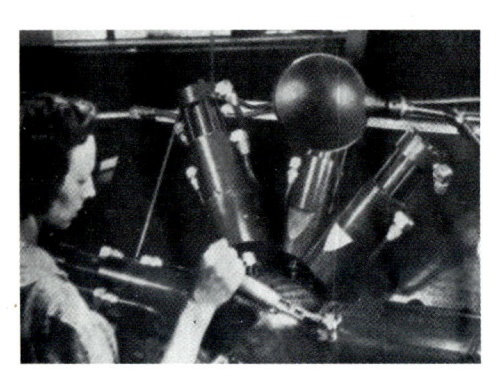

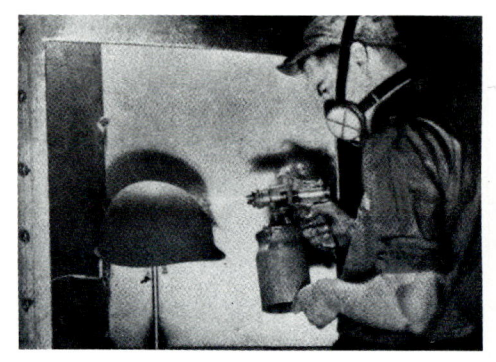

Left: The helmet liner was pierced on the piercing press. The press punched 12 holes into the liner to admit the various rivets and eyelet. (Courtesy of U.S. Army Quartermaster Corps) Center: Various components were riveted to the helmet liner. Separate stations were used to rivet the different components. (Courtesy of U.S. Army Quartermaster Corps) Right: Manually spraying the exterior of the M-1 helmet liner with olive drab paint at the Inland Manufacturing plant, ca. 1942. Some plants contained an automatic spray booth to perform this operation. (Courtesy of GM Media Archives, Copyright 1978 GM Corp. used with permission)

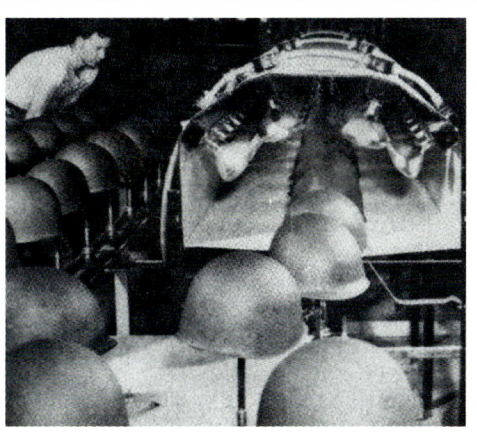

CHAPTER FIVE: PLASTIC HELMET LINER PRODUCTION: 1942-1945

Right: The painted helmet liner was dried in an oven of infra-red lamps. The olive drab paint with resin particles was only applied to the exterior of the helmet. The newly painted liners are shown entering the oven containing the newly invented Westinghouse infra-red lamps at the Westinghouse Micarta Division. (Courtesy of Westinghouse Electric Corporation) Far right: Final inspection on the completed M-1 plastic helmet liner. (Courtesy of U.S. Army Quartermaster Corps)

Below: Unpainted Westinghouse M-1 plastic helmet liner. (Courtesy of Keith R. Jamieson, M.D.)

two slit buckle, which was considered too hard to adjust by the Army. Adjustment was made possible by use of a cam lever, which had an upwardly curled rear resilient section. Mr. Skinner had observed a folding match cover and an airline seat buckle, and used these two items as the basis for the development of the spring clip and chin strap buckle respectively.

With a new chin strap buckle developed, the Army sought development of a new method to secure the leather chin strap to the liner. Previously the chin strap was permanently secured to the liner, but this was to the disliking of the Army which had an already established method of cleaning the helmets. The cleaning procedure was such that high temperatures were used which caused the leather chin strap to be baked to a crisp during the operation. The United Carr Fastener Company developed a new fastener, which consisted

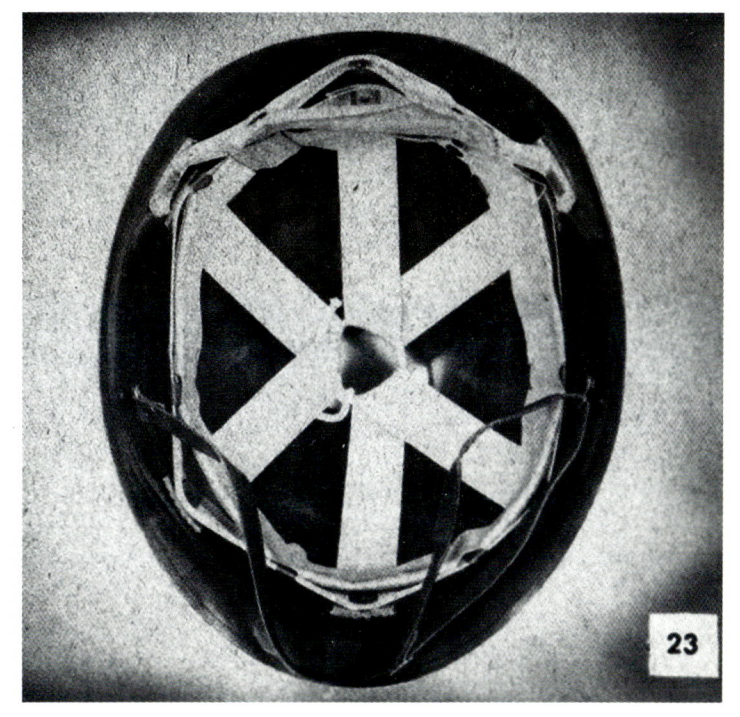

Left: M-1 plastic helmet liner. Exterior side view. This is an early variation of the high-pressure plastic helmet liner. (Courtesy of U.S. Army Ordnance Department via National Archives, RG 156 Ordnance) Right: M-1 plastic helmet liner. Interior view. This is an early variation of the high-pressure plastic helmet liner as it contains the early rayon suspension with perpendicular-sided washers and permanent leather chin strap. Note the head band which is the newly invented adjustable head band made of cotton twill, olive drab shade no. 3, full leather, and spring clips. (Courtesy of U.S. Army Ordnance Department via National Archives, RG 156 Ordnance)

THE M-1 HELMET

Interior view of the Westinghouse M-1 plastic helmet liner, ca. May 1942-September 1942. Components consist of rayon suspension, steel perpendicular-sided washers, and permanent leather chin strap, not shown.

Exterior side view of the early variation M-1 high-pressure plastic helmet liner manufactured by the Westinghouse Electric Company, ca. May 1942-September 1942. Note permanent leather liner chin strap. (Courtesy of Dallas W. Freeborn)

of a garter stud and hook that permitted the chin strap to be securely fastened to the liner and yet removable.

All of these items became standard for production in June 1942, and all the hardware items were to be made of steel painted with a coat of olive drab, due to the ongoing shortages of brass. Although these new items had become standardized for production in June 1942, contract commitments and production change over did not occur until late summer of 1942. Stamp markings often appeared on the new items and webbing. The head band and neck band usually displayed the contract number and the manufacturer for those items. The neck band also included a size number. The introduction of these new items not only affected the plastic helmet liner production but were also used on the remaining fiber liners which did not complete production until November 1942.

By the spring of 1943, a suitable finish had been developed for the plastic helmet liner. The original paint mixture produced plastic helmet liners with a dark olive drab color and a smooth finish. The original paint mixture also did not adhere very well to the helmet and as a result tended to chip off and reveal the shiny laminated surface of the liner. Efforts to solve this problem recommended that the liners be sanded before the painting process was begun, but this method was soon dropped in favor of a new type of paint mixture developed by the Forbes Varnish Company, Cleveland, Ohio. Between October 1942 and March 1943, all of the plastic helmet liner manufacturers had switched to the new paint mixture which contained small particles of phenolic resin and produced a textured finish. The new paint not only adhered better but also had less reflectivity and appeared to be a lighter shade of olive drab.

Also refined in the spring of 1943 was the two piece bar buckle used in the adjustable head band. The new buckle consisted of a single piece of metal, which was similar to a rectangular two slit buckle, except that one slit had rounded teeth on one side, while the

Right: Interior view of the Westinghouse M-1 plastic helmet liner, ca. May 1942-September 1942. Components consist of rayon suspension, steel perpendicular-sided washers, and permanent leather chin strap. (Courtesy of Dallas W. Freeborn)

CHAPTER FIVE: PLASTIC HELMET LINER PRODUCTION: 1942-1945

Left: Interior view of the Mine Safety Appliances M-1 plastic helmet liner, ca. September 1942. Components consist of rayon suspension, unpainted steel A washers, and permanent leather chin strap. (Courtesy of Keith R. Jamieson, M.D.) Center: Interior view of Westinghouse M-1 plastic helmet liner with rayon suspension and steel unpainted A washers, ca. September 1942. (Courtesy of Keith R. Jamieson, M.D.) Right: Interior view of M-1 plastic helmet liner, ca. September 1942-early 1943. This particular liner was manufactured by Firestone. Components included cotton olive drab shade no. 3 suspension and neck strap, unpainted steel A washers, and a removable leather chin strap.

other slit was completely smooth. In the fall of 1943, the chin strap buckle clip design was also modified from the straight edge to a roller, to allow any thickness of leather chin strap to be admitted. The modified chin strap buckle was later patented by the American Fastener Company, Waterbury, Connecticut, and remained the only known patented helmet item to come out of the entire M-1 Helmet development program during World War II.

The number of companies producing plastic helmet liners was reduced from nine to six during the early part of 1944. Two of the companies, which did not get contracts for additional liners, were producing types of helmet liners that the Army felt they no longer needed. The St. Clair low-pressure helmet liner, which was initially purchased to fill the requirement void for plastic helmets while the high-pressure production method was refined, and the Hood ball-winding method were now considered unsatisfactory by the Army. The third company not to receive a new contract was the Inland Manufacturing Division of General Motors. The Army was in no way unsatisfied with the quality of their work, but the company was simply located in an area where the work force was needed in other fields of production.

Shortly after the number of contracts was reduced, the U.S. Army Quartermaster recommended that brass be adopted for use in the manufacture of all hardware items for the helmet liner, with the exception of the spring clips. The recommendation was made in March 1944, at which time brass was no longer at critical shortage levels, and production was switched to brass by June 1944. The brass used in the hardware of the liner was coated with a rust and mildew inhibitor that gave the brass items a flat black coat like finish. Other changes in the specification during the spring of 1944, allowed for the use of double and triple Herringbone twill weave, instead of the single Herringbone twill weave, in the production of the khaki, olive drab shade no. 3, cotton webbing, and the neck band's size number was replaced by a stamp with the word small, medium, or large.

In September 1944, after the completion of an extensive study at Camp Indian Bay, Florida, by the Quartermaster Board, a suggestion was made that the sized neck band be replaced by a single adjustable neck band that would fit all sizes. Shortly thereafter, action was taken and a single adjustable neck band was manufactured and issued. This was the last major change made on the liner before the end of the war.

PLASTIC HELMET LINER TOTALS

Ten companies produced the helmet liner shell, while an additional 30 were responsible for the manufacture of the various components that went into the liner. The total number of liners produced, including the 3,977,000 fiber liners, 1,506,000 low-pressure plastic liners, and the approximately 38,217,000 high-pressure plastic liners, were at least 43,700,000 during the years between 1941-1945. Helmet liner production was discontinued around 17 August 1945, just two days after VJ-Day. Westinghouse was the largest manufacturer of the helmet liner, accounting for about half of the liners produced, with its Micarta Division producing approximately 13,000,000 and its Bryant Electric Division producing approximately 10,000,000. Firestone produced approximately 7,500,000 plastic liners, and the Inland Manufacturing Division, which stopped producing helmet liners in 1943, completed approximately 1,900,000 liners. The remaining companies, Mine Safety, Capac, Seaman, and International Molded Plastics each produced between 2,000,000 and 4,000,000 plastic helmet liners.

THE M-1 HELMET

Left: M-1 helmet liner A washer. This early variation of the A washer was made of steel and left unpainted, ca. 1942. Right: M-1 helmet liner adjustable head band. This early variation of the adjustable head band consisted of single Herringbone Twill cotton webbing in olive drab shade no. 3, with leather, spring clips, and early two piece steel bar buckle, ca. 1942.

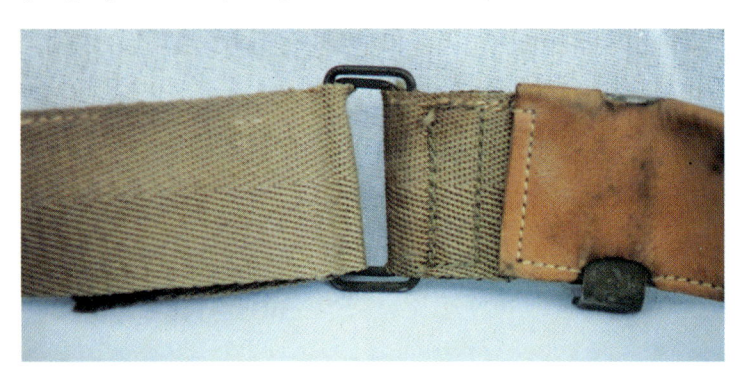

Left: Front view of the M-1 helmet liner head band bar buckle with head band secured. This is the early variation of the bar buckle consisting of two separate pieces and made of steel painted olive drab, ca. 1942-1943. Right: Rear view of the M-1 helmet liner head band bar buckle with head band secured. This view clearly shows the two separate pieces.

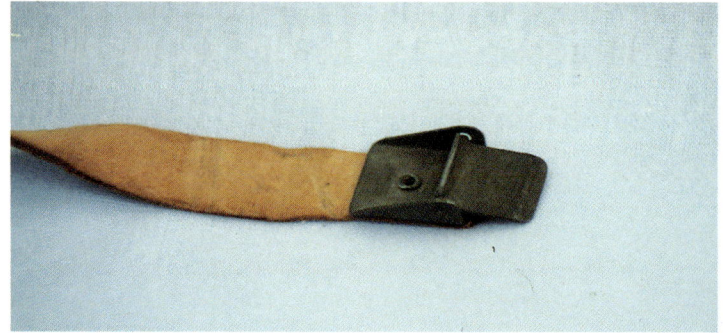

Left: M-1 helmet liner chin strap, ca. 1942-1943. Right: M-1 helmet liner chin strap wedge buckle. This early variation of the chin strap buckle is identified by its straight edge and is made of steel painted olive drab, ca. 1942-1943.

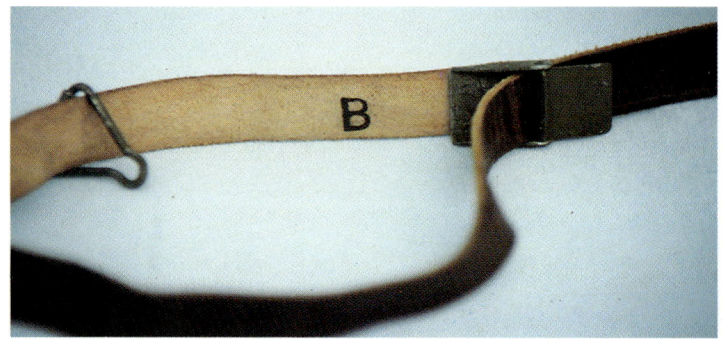

Left: M-1 helmet liner chin strap wedge buckle. This early variation of the chin strap buckle is identified by its straight edge and is made of steel painted olive drab, ca. 1942-1943. Right: Close-up of the M-1 helmet liner chin strap with stamp marking. This is an early variation of the chin strap and the B marking represents its manufacturer, Belvidere, Inc., Brooklyn, New York, ca. 1943-1944.

CHAPTER FIVE: PLASTIC HELMET LINER PRODUCTION: 1942-1945

Left: M-1 helmet liner head band spring clip. This early variation of the spring clip is made of steel painted olive drab, ca. 1942-1943. Right: M-1 helmet liner chin strap holder, also called a garter hook. This variation of the chin strap holder is made of steel painted olive drab, ca. 1942-1944.

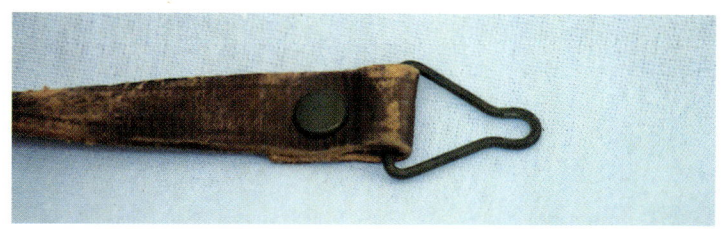

Left: M-1 helmet liner chin strap holder with chin strap. This early variation of the chin strap holder is made of steel painted olive drab, and the rivet made of steel painted brown, ca. 1942-1943. Right: M-1 helmet liner chin strap holder with chin strap. This variation of the chin strap holder is made of steel painted olive drab, ca. 1943-1945.

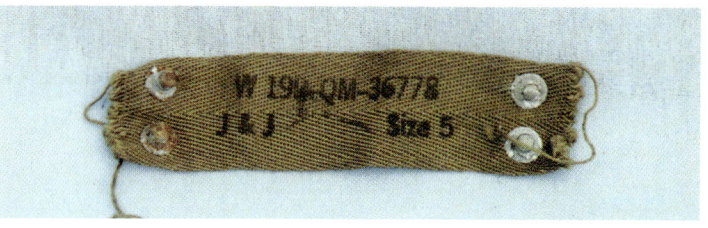

Left: Close-up of the "anchor" stamp on the M-1 helmet liner chin strap buckle, ca. 1942-1944. Right: M-1 helmet liner neck band, size 5. This early variation of the cotton neck band was contracted by the Johnson & Johnson Company during July 1942.

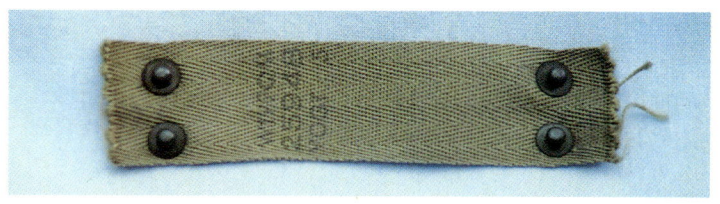

Left: M-1 helmet liner neck band, contracted by the Vogt Manufacturing Corporation during June 1942. This variation of the cotton neck band was made using triple Herringbone Twill cotton. Triple Herringbone Twill was allowed as a substitute for single Herringbone Twill. Right: M-1 helmet liner neck band, ca. 1942. This variation of the cotton neck band was made using triple Herringbone Twill cotton. The neck band was made under the same contract as the one shown earlier, but was manufactured with brass snap fasteners.

Left: M-1 helmet liner chin strap garter stud, ca. 1942-1944. This early variation of the garter stud was made of steel. Note the dimpled top in this design. Center: M-1 helmet liner chin strap garter stud, ca. 1942-1944. This early variation of the garter stud was made of steel. Note the smooth top in this design. Right: M-1 helmet liner suspension snap fastener socket, ca. 1942-1944. This variation of the snap fastener socket was made of steel.

THE M-1 HELMET

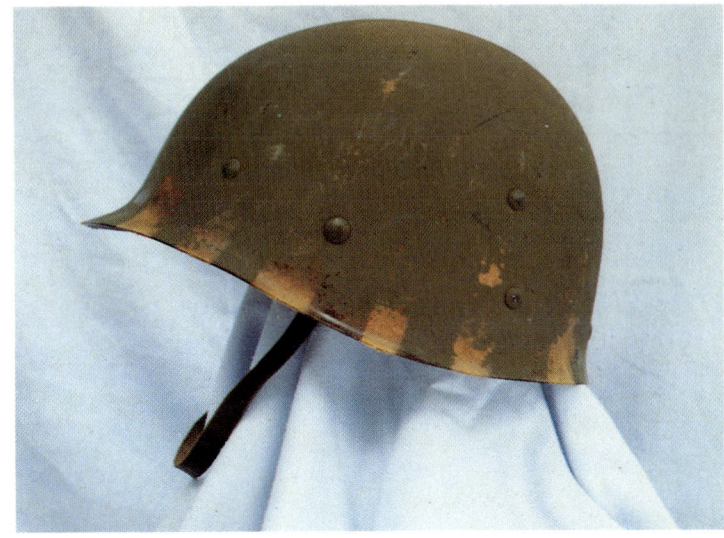

Left: Early and late shades of olive drab. The Westinghouse helmet on the left has the earlier dark olive drab paint job with no resin texturing, while a later manufactured Firestone M-1 helmet liner has the early variation of the lighter olive drab paint job with resin texture. This early paint with resin texture tended to chip quite a bit, as can be seen. Right: Exterior right side view of the Mine Safety M-1 plastic helmet liner, early 1943. This early plastic helmet liner has the early light olive paint job with resin texture. The larger rivet on the side indicates that this helmet has the removable leather chin strap.

Left: Interior of the M-1 plastic helmet liner complete, ca. early 1943. This particular plastic helmet liner was manufactured by the Mine Safety Appliances Company. Right: Interior view of M-1 plastic helmet liner, ca. early 1943-June 1944. Components included cotton olive drab shade no. 3 suspension and neck strap, steel A washers painted olive drab, and a removable leather chin strap. Note the P.K.W. stamp near the front part of the liner suspension. P.K.W. represents the Paul K. Weil Company, St. Louis, Missouri.

Left: M-1 helmet liner A washer. This variation of the A washer is made of steel painted olive drab, ca. 1943-1944. Right: M-1 helmet liner suspension with stamp marking. The stamp marking, J&J, represented the Johnson & Johnson Company.

50

CHAPTER FIVE: PLASTIC HELMET LINER PRODUCTION: 1942-1945

Left: M-1 helmet liner adjustable head band. This variation of the adjustable head band consists of single Herringbone Twill cotton webbing in olive drab shade no. 3, with leather, spring clips, and single piece stamped steel bar buckle, ca. 1943-1944. Right: M-1 helmet liner head band bar buckle. This later variation of the bar buckle is made from a single piece of stamped steel painted olive drab.

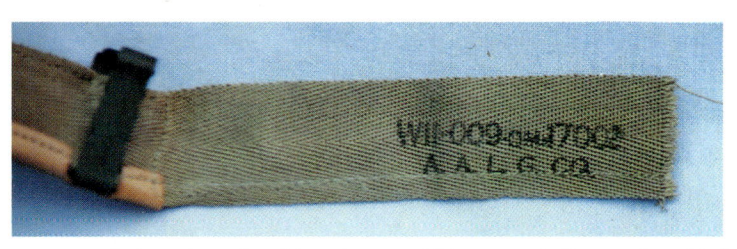

Left: M-1 helmet liner head band stamp. This stamp indicates that this head band was contracted by the All American Leather Goods Company of Chicago Illinois during November 1943. Right: M-1 helmet liner head band spring clip. This later variation of the spring clip is made of steel painted olive drab, ca. 1943-1945.

Left: M-1 helmet liner chin strap wedge buckle. This variation of the chin strap buckle is identified by its rolled edge and is made of steel painted olive drab, ca. 1943-1944. Right: Close-up of the M-1 helmet liner chin strap with stamp marking. This chin strap with the GB marking represents Goldsmith Brothers & Manufacturing Company, New York, New York, ca. 1943-1944.

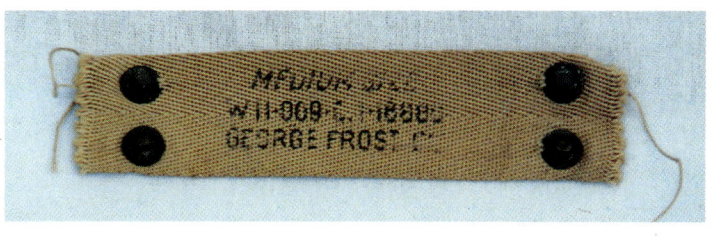

Left: M-1 helmet liner neck band, size small. This variation of the neck band was made of single Herringbone Twill cotton in olive drab shade no. 3 with steel male snap fasteners by the American Stay Company. The number under the company name is the contract number. This particular contract, W 199 QM 33826, was issued to the American Stay Company on 5 May 1943, and was for 800,000 neck bands at less than 2 cents apiece. Right: M-1 helmet liner neck band, size medium. This variation of the neck band was made using brass male snap fasteners and contracted to the George Frost Company during October 1943.

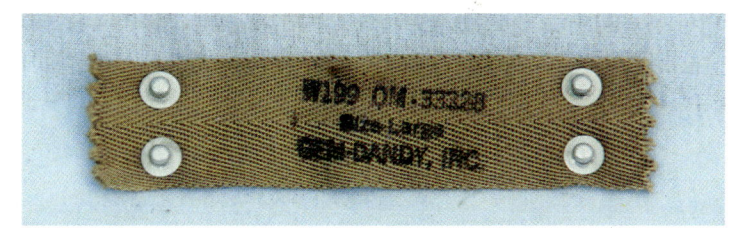

M-1 helmet liner neck band, size large. This neck band was contracted to Gem Dandy, Inc. during May 1943.

THE M-1 HELMET

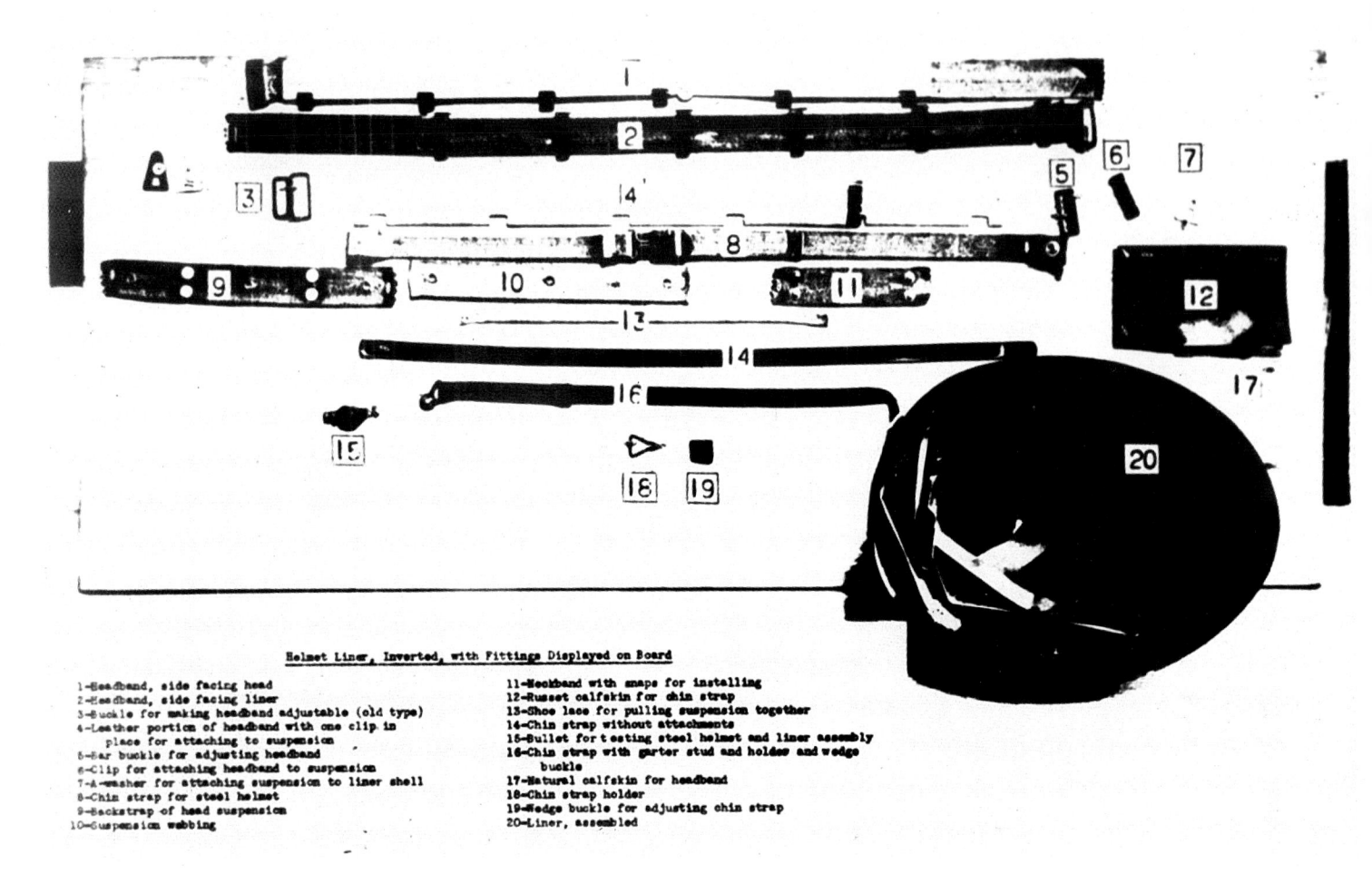

Helmet Liner, Inverted, with Fittings Displayed on Board

1—Headband, side facing head
2—Headband, side facing liner
3—Buckle for making headband adjustable (old type)
4—Leather portion of headband with one clip in place for attaching to suspension
5—Bar buckle for adjusting headband
6—Clip for attaching headband to suspension
7—A-washer for attaching suspension to liner shell
8—Chin strap for steel helmet
9—Backstrap of head suspension
10—Suspension webbing
11—Neckband with snaps for installing
12—Russet calfskin for chin strap
13—Shoe lace for pulling suspension together
14—Chin strap without attachments
15—Bullet for testing steel helmet and liner assembly
16—Chin strap with garter stud and holder and wedge buckle
17—Natural calfskin for headband
18—Chin strap holder
19—Wedge buckle for adjusting chin strap
20—Liner, assembled

M-1 helmet liner display board. (Courtesy of U.S. Army Quartermaster Corps)

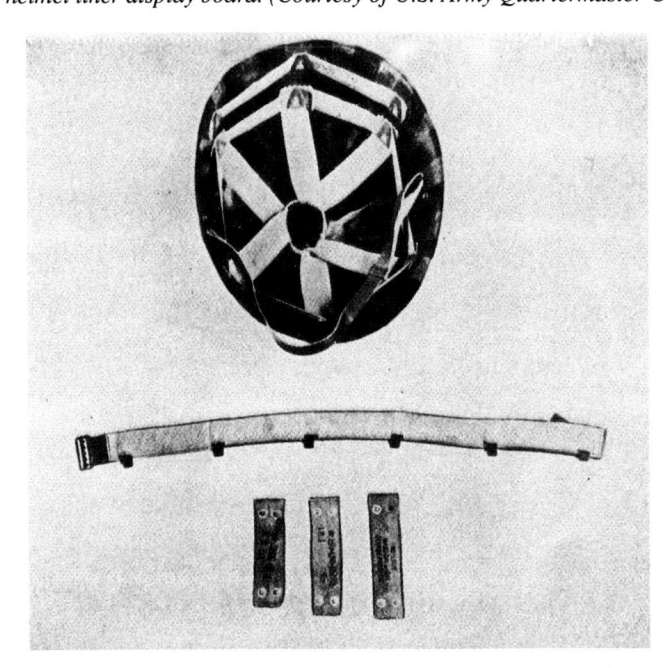

M-1 plastic helmet liner. Top, interior view of the plastic liner showing assembled cotton suspension and removable leather chin strap. This was how the liner appeared when shipped from the factory. Below, the additional items were issued when the helmet was issued to the soldier and included the head band and neck band. The head band shown here was the adjustable cotton head band with one piece bar buckle, sewn full leather, and spring clips. The neck band shown here was the cotton sized neck band with sizing numbers. (Courtesy of U.S. Army Quartermaster Corps, Library Collections of the U.S. Army Quartermaster Museum, Fort Lee, Virginia)

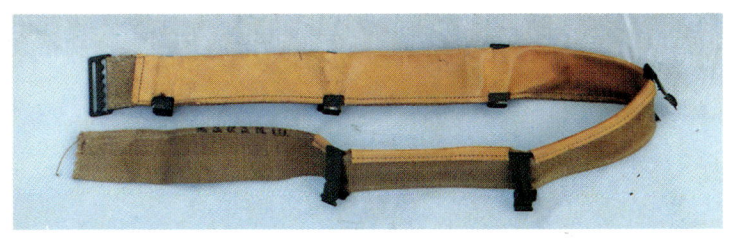

CHAPTER FIVE: PLASTIC HELMET LINER PRODUCTION: 1942-1945

Left: M-1 helmet liner adjustable head band. This variation of the adjustable head band consists of single Herringbone Twill cotton webbing in olive drab shade no. 3, with leather, spring clips, and single piece stamped steel bar buckle, ca. 1943-1944. Right: M-1 helmet liner head band bar buckle. This later variation of the bar buckle is made from a single piece of stamped steel painted olive drab.

Left: M-1 helmet liner head band stamp. This stamp indicates that this head band was contracted by the All American Leather Goods Company of Chicago Illinois during November 1943. Right: M-1 helmet liner head band spring clip. This later variation of the spring clip is made of steel painted olive drab, ca. 1943-1945.

Left: M-1 helmet liner chin strap wedge buckle. This variation of the chin strap buckle is identified by its rolled edge and is made of steel painted olive drab, ca. 1943-1944. Right: Close-up of the M-1 helmet liner chin strap with stamp marking. This chin strap with the GB marking represents Goldsmith Brothers & Manufacturing Company, New York, New York, ca. 1943-1944.

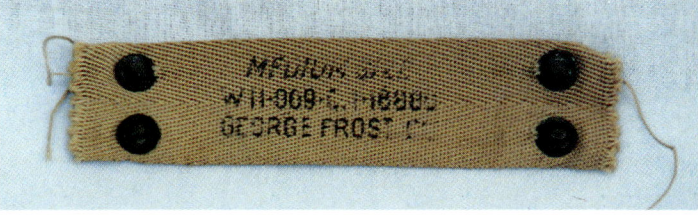

Left: M-1 helmet liner neck band, size small. This variation of the neck band was made of single Herringbone Twill cotton in olive drab shade no. 3 with steel male snap fasteners by the American Stay Company. The number under the company name is the contract number. This particular contract, W 199 QM 33826, was issued to the American Stay Company on 5 May 1943, and was for 800,000 neck bands at less than 2 cents apiece. Right: M-1 helmet liner neck band, size medium. This variation of the neck band was made using brass male snap fasteners and contracted to the George Frost Company during October 1943.

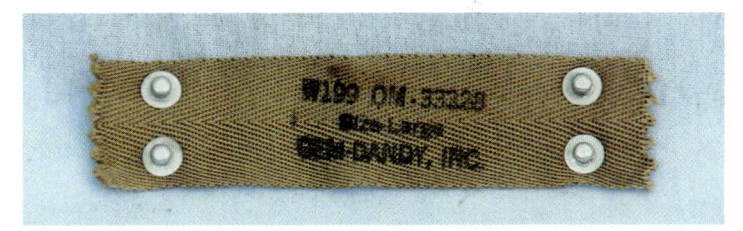

M-1 helmet liner neck band, size large. This neck band was contracted to Gem Dandy, Inc. during May 1943.

51

THE M-1 HELMET

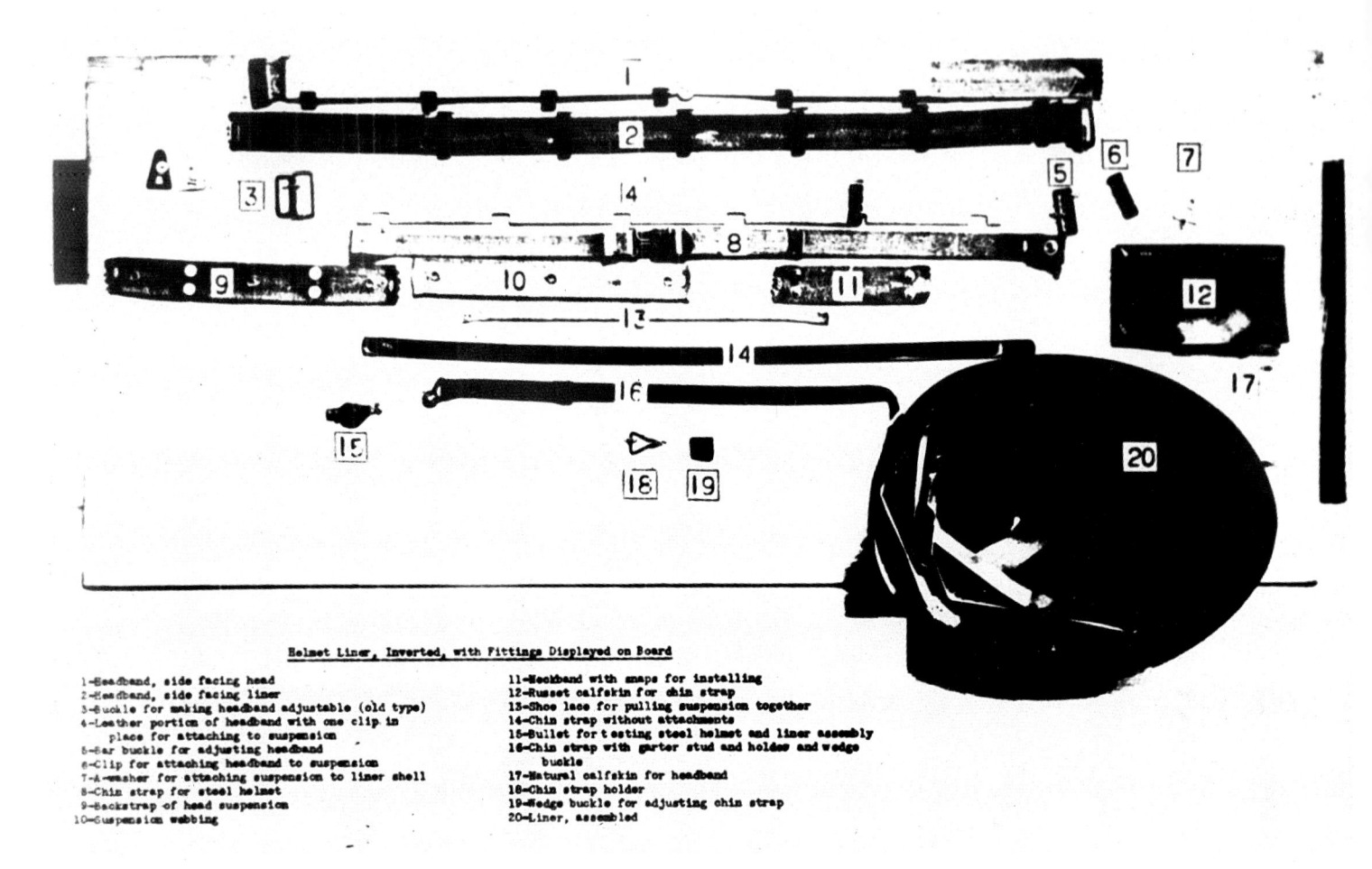

Helmet Liner, Inverted, with Fittings Displayed on Board

1—Headband, side facing head
2—Headband, side facing liner
3—Buckle for making headband adjustable (old type)
4—Leather portion of headband with one clip in place for attaching to suspension
5—Bar buckle for adjusting headband
6—Clip for attaching headband to suspension
7—A-washer for attaching suspension to liner shell
8—Chin strap for steel helmet
9—Backstrap of head suspension
10—Suspension webbing
11—Neckband with snaps for installing
12—Russet calfskin for chin strap
13—Shoe lace for pulling suspension together
14—Chin strap without attachments
15—Bullet for testing steel helmet and liner assembly
16—Chin strap with garter stud and holder and wedge buckle
17—Natural calfskin for headband
18—Chin strap holder
19—Wedge buckle for adjusting chin strap
20—Liner, assembled

M-1 helmet liner display board. (Courtesy of U.S. Army Quartermaster Corps)

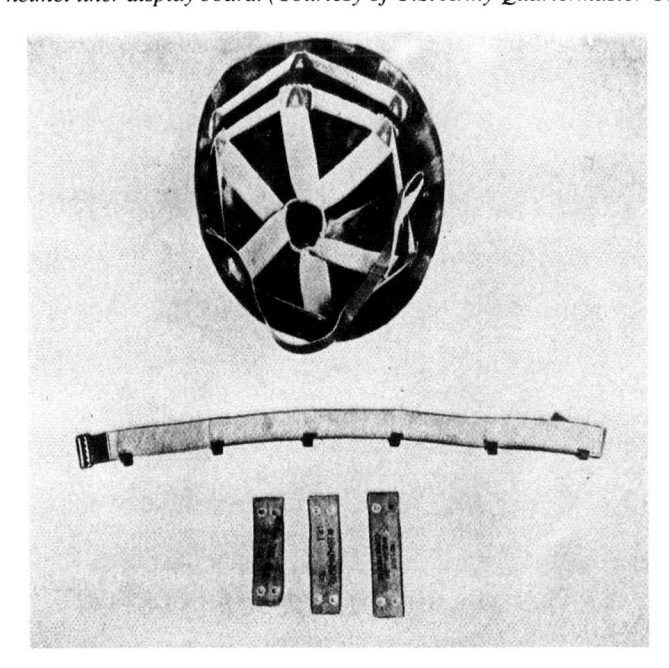

M-1 plastic helmet liner. Top, interior view of the plastic liner showing assembled cotton suspension and removable leather chin strap. This was how the liner appeared when shipped from the factory. Below, the additional items were issued when the helmet was issued to the soldier and included the head band and neck band. The head band shown here was the adjustable cotton head band with one piece bar buckle, sewn full leather, and spring clips. The neck band shown here was the cotton sized neck band with sizing numbers. (Courtesy of U.S. Army Quartermaster Corps, Library Collections of the U.S. Army Quartermaster Museum, Fort Lee, Virginia)

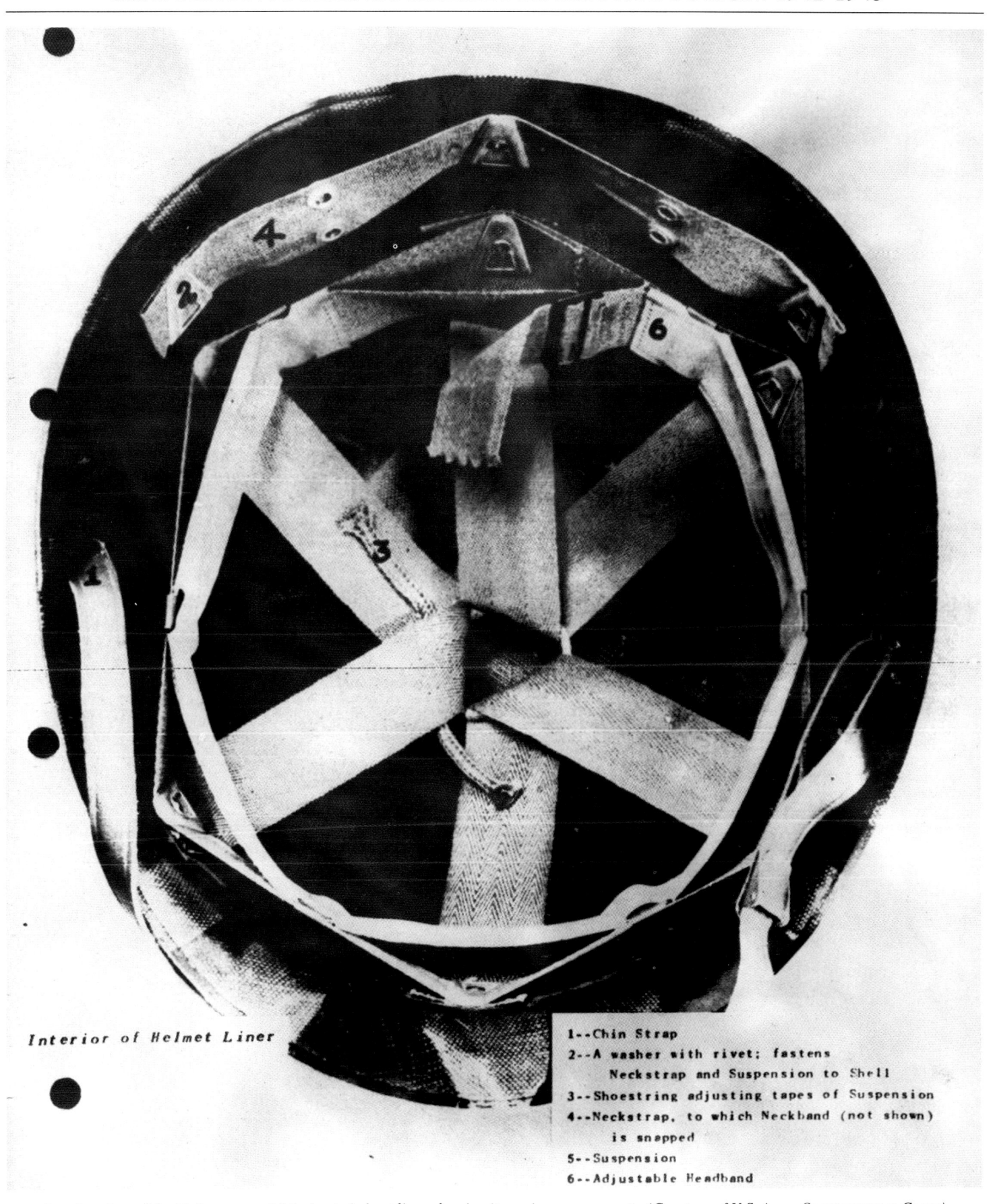

CHAPTER FIVE: PLASTIC HELMET LINER PRODUCTION: 1942-1945

Interior of Helmet Liner

1 -- Chin Strap
2 -- A washer with rivet; fastens Neckstrap and Suspension to Shell
3 -- Shoestring adjusting tapes of Suspension
4 -- Neckstrap, to which Neckband (not shown) is snapped
5 -- Suspension
6 -- Adjustable Headband

Interior view of the high-pressure M-1 plastic helmet liner showing its various components. (Courtesy of U.S. Army Quartermaster Corps)

THE M-1 HELMET

Left: Interior view of M-1 plastic helmet liner, ca. June 1944-August 1945. This particular liner was manufactured by Westinghouse. Components included cotton olive drab shade no. 3 suspension and neck strap, brass A washers coated in a mildew inhibitor, and a removable leather chin strap. Right: Exterior view of the Westinghouse M-1 plastic helmet liner, ca. June 1944-August 1945.

Left: M-1 helmet liner A washer. This late variation of the A washer is made of brass and coated with a mildew inhibitor, ca. June 1944-August 1945. Center: M-1 helmet liner chin strap garter stud, ca. 1944-1945. This late variation of the garter stud was made of brass and coated with a mildew inhibitor. Right: M-1 helmet liner suspension snap fastener socket, ca. 1944-1945. This variation of the snap fastener socket was made of brass and coated with a mildew inhibitor. This particular fastener was manufactured by the United Carr Fastener Company.

Left: M-1 helmet liner chin strap wedge buckle. This later variation of the chin strap buckle is identified by its rolled edge and is made of brass coated with a mildew inhibitor, ca. 1944-1945. Right: Close-up of the patent number stamp on the M-1 helmet liner chin strap buckle, ca. June 1944-August 1945. The patent, 2363872, was applied for by the American Fastener Company, Waterbury Connecticut in March 1943, and the patent was granted in November 1944.

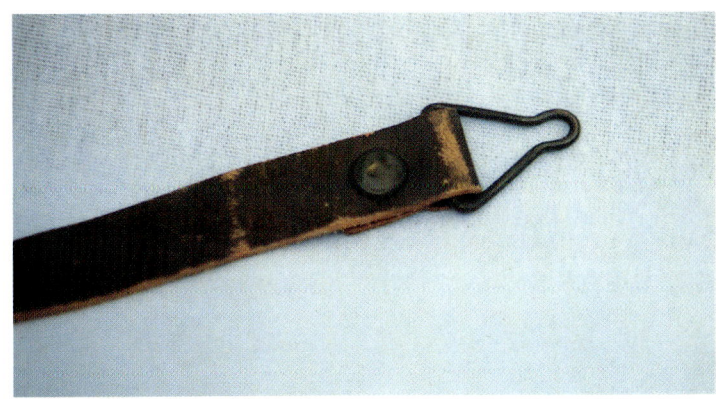

CHAPTER FIVE: PLASTIC HELMET LINER PRODUCTION: 1942-1945

Left: Close-up of the M-1 helmet liner chin strap with stamp marking. This is a later variation of the chin strap and the H marking represents The Hagerstown Leather Company, Hagerstown, Maryland, ca. 1944-1945. Right: Front view of the M-1 helmet liner chin strap holder with chin strap. This later variation of the chin strap holder is made of brass coated with a mildew inhibitor, ca. 1944-1945.

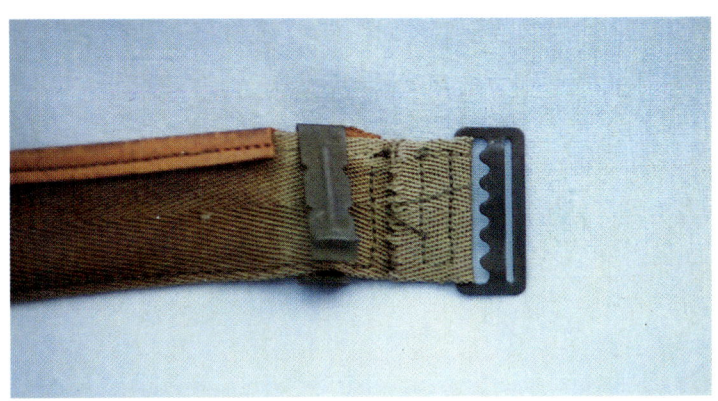

Left: Rear view of the M-1 helmet liner chin strap holder with chin strap. This later variation of the chin strap holder is made of brass coated with a mildew inhibitor, ca. 1944-1945. Right: M-1 helmet liner head band stamped buckle, ca. June 1944-August 1945. This variation of the buckle was made of brass coated with a mildew inhibitor.

M-1 helmet liner adjustable neck band. The adjustable neck band was made of single Herringbone Twill cotton in olive drab shade no. 3, brass male snap fasteners, and a stamped brass bar buckle. This particular adjustable neck band was made by the American Stay Company during early 1945. The number above the company name is the specification number. This particular specification, C.Q.D. 64-B, was issued by the Chicago Quartermaster Depot as the 64th item, revision B.

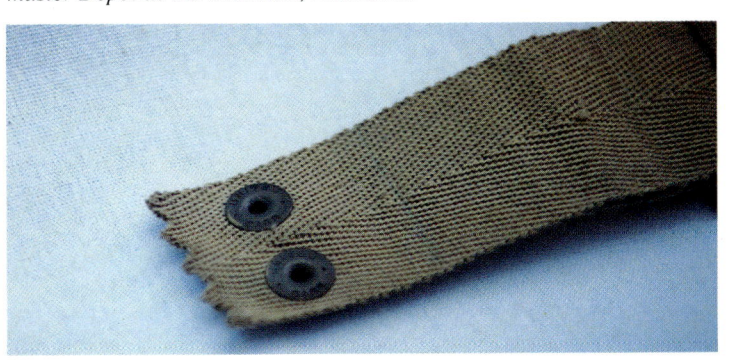

Above: M-1 helmet liner. This picture shows a U.S. Army soldier modeling an Inland Manufacturing Division high-pressure M-1 plastic helmet liner, ca. 1942. (Courtesy of GM Media Archives, Copyright 1978 GM Corp. used with permission)

Left: Rear view of the brass snap fastener manufactured by the United Carr Fastener Company. The brass fastener is coated with a mildew inhibitor.

THE M-1 HELMET

Left: M-1 plastic helmet liner crown stamp. This crown stamp, the word INLAND within a crown, identifies the manufacturer as the Inland Manufacturing Division of General Motors. Center: M-1 plastic helmet liner crown stamp. This crown stamp, the letter W on an oval, both enclosed in a circle, identifies the manufacturer as the Westinghouse Electric Company. Right: M-1 plastic helmet liner crown stamp. This crown stamp, stylistic letters MSA in a circle, identifies the manufacturer as The Mine Safety Appliances Company.

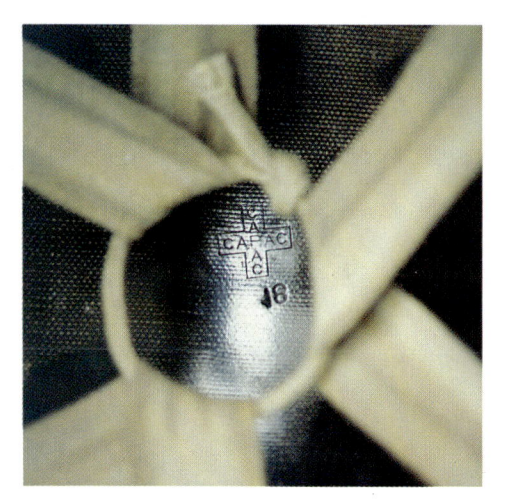

Left: M-1 plastic helmet liner crown stamp. This crown stamp, the symmetric letters CAPAC in a cross, identifies the manufacturer as the Capac Manufacturing Company. Right: M-1 plastic helmet liner crown stamp. This crown stamp, a stylistic letter F in a shield, identifies the manufacturer as the Firestone Tire & Rubber Company.

Above: The M-1 plastic helmet liner worn as a field hat behind the front lines by Raymond Reynosa during a break in the Battle of the Bulge, winter 1944-1945.

Far left: M-1 plastic helmet liner crown stamp. This crown stamp, a small man with arms on hips and wearing a crown and the letters IMP below him, identifies the manufacturer as the International Molded Plastics Company.

Left: M-1 plastic helmet liner crown stamp. This crown stamp, the letter S in an oval, identifies the manufacturer as the Seaman Paper Company.

CHAPTER FIVE: PLASTIC HELMET LINER PRODUCTION: 1942-1945

The plastic helmet liner was often worn as a field hat behind the front lines. Here it is worn by soldiers in Europe during winter 1944-1945. (Courtesy of U.S. Army)

A Marine Corps chow line showing soldiers wearing both the M-1 fiber liner, at left with dimples around the rivets, and the M-1 plastic liner, at right. (Courtesy of U.S. Marine Corps)

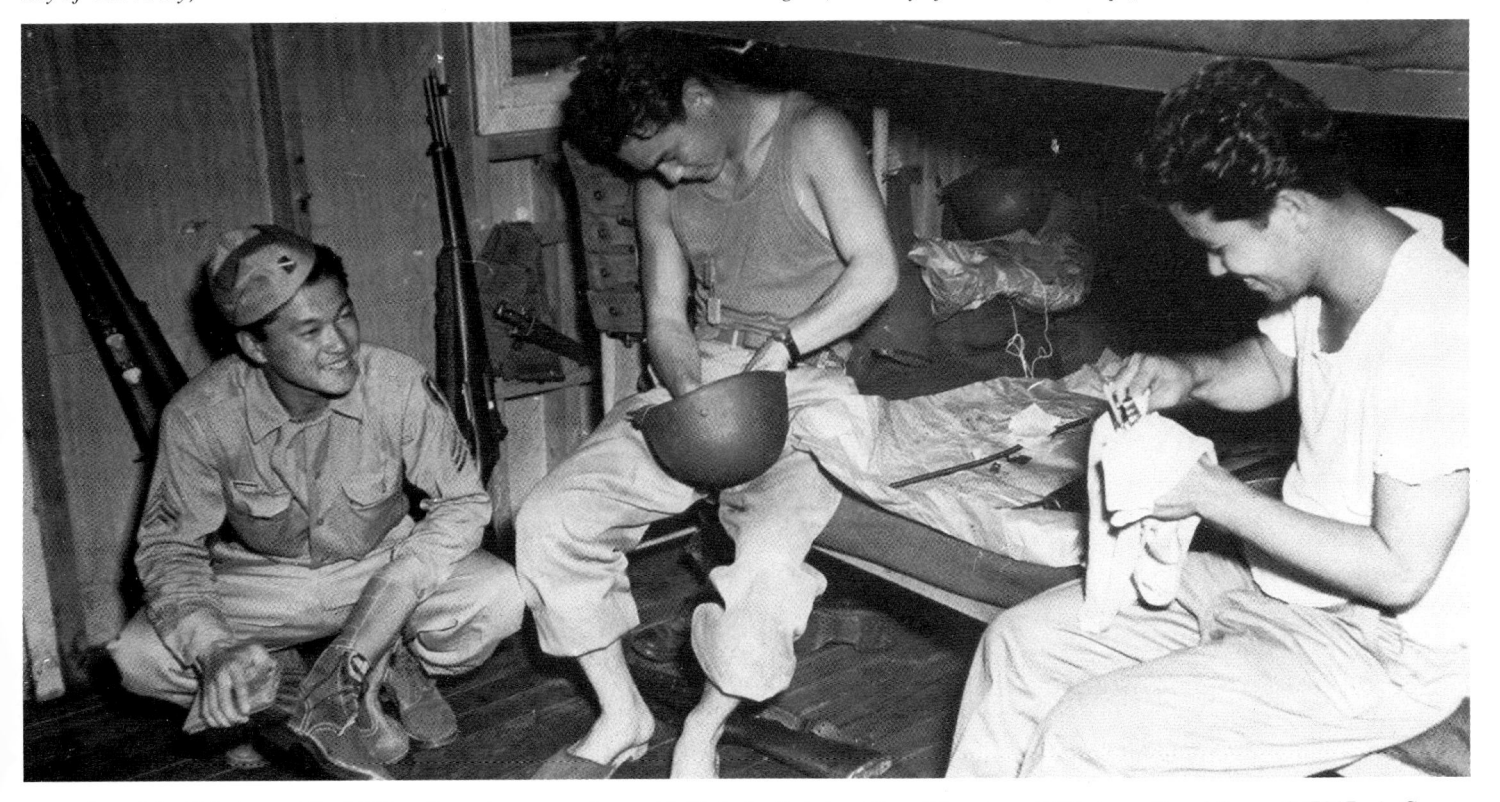

A member of the U.S. Army 442nd CIR, adjusts the suspension on his M-1 plastic helmet liner. (Courtesy of U.S. Army Signal Corps, 247017, via George A. Petersen)

CHAPTER SIX

PROCUREMENT, DISTRIBUTION, AND THE COMPLETE HELMET

In 1941, the McCord Radiator and Manufacturing Company had received contracts for the entire M-1 Helmet, and in turn they sub-contracted for most of the helmet's component parts. McCord was responsible for delivering the completed M-1 Helmets to the U.S. Army.

With the development of the plastic liner, and later the almost entirely new suspension system, the Ordnance Department and the Quartermaster Corps decided to break up the single contracts for the M-1 Helmets, as had been previously given to the McCord Radiator and Manufacturing Company. By the spring of 1942, the Ordnance Department was only issuing contracts for the helmet body, while the Quartermaster Corps issued separate contracts for the liner body, the head band and the neck band.

Since the various helmet components were contracted for separately, it was probable that the Quartermaster ordered adjustable head bands and neck bands in excess of the contracted helmet liners. The extra head bands and neck bands likely served as spares. Spares for the leather liner chin strap were contracted for, since it was not considered a separate component, but part of the liner when initially procured. At least 2,588,765 leather liner chin straps were contracted for as spares. This was rather a low number, attributed to the Quartermaster's awareness that most of the wearers did not actually use the chin strap, but often placed them over the brim of the helmet.

Since the M-1 helmet was not manufactured as a complete unit, it was necessary to have Distribution Depots which would bring the four major components of the helmet together for issuing. The four major components included the steel helmet body, helmet liner, head band, and neck band. Once together at the Depots, which included Atlanta, Kansas City, Chicago, Memphis, Richmond, and Schenectady, the helmet components could then be requested by camps within the United States or overseas to meet their require-

Glider troops wearing the early M-1 helmet body with the M-1 Fiber helmet liner. Note the highly visible thick rim of fiber liner. (Courtesy of U.S. Army)

CHAPTER SIX: PROCUREMENT, DISTRIBUTION, AND THE COMPLETE HELMET

Exterior and interior views of the Hawley M-1 fiber helmet liner with M-1 steel body, ca. late 1941-May 1942. The fiber liner was quite noticeable when worn with the M-1 steel body.

ments. The helmets were then usually issued to troops at various reception centers. Initially, packaging of the helmet and its various components for overseas, was accomplished by just a few firms, such as Westinghouse and Firestone. Later, due to the large number of overseas request, all the companies manufacturing components for the helmet were required to package their own items for shipment and distribution.

Procurement of the M-1 Helmet by the U.S. Navy and the U.S. Marine Corps were handled as sale directives through the U.S. Army, which distributed the helmets to the requesting service.

The cost of the complete M-1 helmet was about $3.00, with the helmet body ranging in cost from $0.83 to $1.03, the helmet liner from $1.25 to $ 2.35, the head band from $0.20 to $0.35, and the neck band from $0.01 to $0.02.

M-1 helmet body and plastic helmet liner. This picture shows the method of slip fitting the helmet body to the plastic helmet liner. (Courtesy of U.S. Army Signal Corps)

The M-1 helmet. The plastic helmet liner is shown being placed in the steel helmet body. (Courtesy of Westinghouse Electric Corporation)

THE M-1 HELMET

Exterior right side view of M-1 helmet complete, ca. early 1943. (Courtesy of Keith R. Jamieson, M.D.)

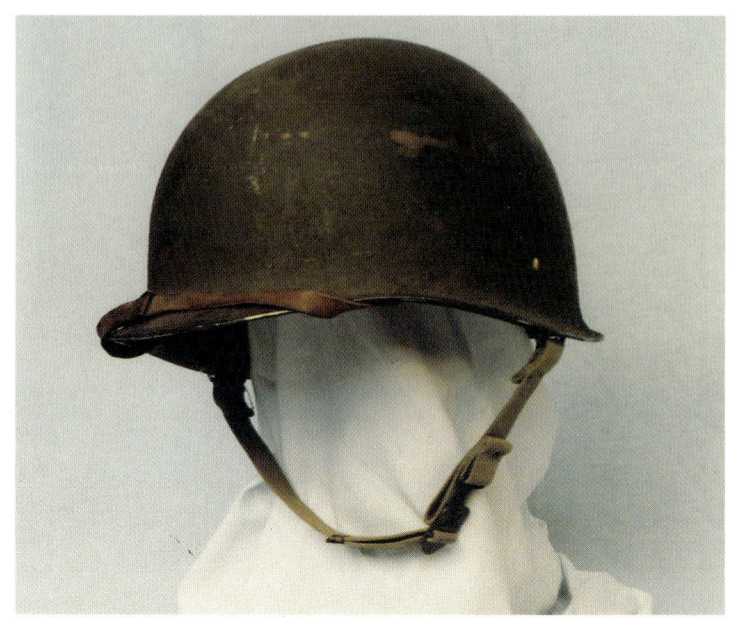

Exterior front left view of M-1 helmet complete, ca. early 1943. (Courtesy of Keith R. Jamieson, M.D.)

Sergeant Harold Gary wears the M-1 helmet complete while assigned to a Sherman tank unit. The stainless steel edging was very prominent on the M-1 helmet, as seen in this picture. Also note that the helmet body's web chin strap has been secured over the rear rim. (Courtesy of U.S Army Signal Corps)

CHAPTER SIX: PROCUREMENT, DISTRIBUTION, AND THE COMPLETE HELMET

Three Marine communications men wear their M-1 helmets. The two outer Marines wear the M-1 helmet complete, while the center Marine wears the M-1 plastic helmet liner. (Courtesy of U.S. Marine Corps)

The M-1 helmet complete worn in the the field. (Courtesy of U.S. Army Signal Corps, 175716, via George A. Petersen)

THE M-1 HELMET

The M-1 helmet complete worn by an anti-aircraft artillery crew. (Courtesy of National Archives, 111-C-1127)

CHAPTER SIX: PROCUREMENT, DISTRIBUTION, AND THE COMPLETE HELMET

M-1 helmet complete, ca. early1943-October 1943.

Soldiers play a game of darts prior to the 6 June 1944 invasion. The soldier on the left wears an early version of the plastic helmet liner with the dark olive drab finish, while the soldier on the right wears the later version plastic helmet liner with the light olive drab finish. (Courtesy of National Archives, 111-C-1123)

USAAF airmen of the 9th Air Force wear various M-1 helmets while relaxing in Paris. The airman on the left wear the Hawley fiber liner, the two airmen in the center wear the complete helmet, and the airman on the right wears the plastic helmet liner. (Courtesy of National Archives, 111-C-2194)

THE M-1 HELMET

U.S. Army PFC Eugene Collins washing up in a fox hole while in training stateside. Clearly visible in this picture are both parts of the M-1 helmet. On the left is the mid-war production version helmet body with hinged flexible chin strap loops. On the right is the mid-war production plastic helmet liner. (Courtesy of National Archives, 111-C-3283)

An infantryman training at Fort Benning and wearing the last variation of the M-1 helmet during May 1945. Note the helmet body's olive drab shade no. 7 web chin strap and manganese rim. (Courtesy of National Archives, 111-C-3316)

Left to right, Alayius J. Kaminski, Raymond J. Luliski and Raymond G. Reynosa behind the lines in Normandy wearing their M-1 helmets and liners during the summer of 1944.

CHAPTER SIX: PROCUREMENT, DISTRIBUTION, AND THE COMPLETE HELMET

Left: M-1 helmet complete, ca. October 1943-November 1944. Right: Interior view of the M-1 helmet complete, ca. October 1943-November 1944.

Exterior view of the M-1 helmet complete, ca. November 1944-August 1945.   *Interior view of the M-1 helmet complete, ca. November 1944-August 1945.*

CHAPTER SEVEN

PARACHUTIST HELMET PRODUCTION AND MODIFICATION: 1942-1945

The concept of a separate helmet for parachutists was considered as early as 1941. By 1942, the United States Army had introduced its first combat paratrooper units. As with other fighting units, the paratroopers required the necessary headgear to carry out their operations. For this role, the standard M-1 helmet was initially tested, but it was found unsuitable. The standard M-1 helmet had a tendency to be knocked off when the parachute opened or when the parachutist assumed an inverted position. To overcome these problems, the Research and Development Branch of the Quartermaster Corps, along with other departments and private industry, developed the necessary modifications for the M-1 helmet to make it suitable for use by parachute troops.

M-2 PARACHUTIST HELMET

The modified M-1 helmet adopted by paratroops differed from the standard M-1 helmet assembly in that the steel body's web chin strap was extended and incorporated a male snap fastener, which allowed the body to be secured to the liner during the jump. This

Experimental TS-3 helmet body and fiber liner shown with early parachutist modifications at Fort Benning, Georgia, during April 1941. The TS-3 helmet would soon be standardized as the M-1 helmet. This photograph shows the fiber liner with additional snap fastener at sides, and the helmet body with pre-production riveted web chin strap and chin strap extensions. (Courtesy of National Archives, 111-SC-118699)

CHAPTER SEVEN: PARACHUTIST HELMET

Sergeant Hart of the U.S. Army paratrooper forces wears a paratooper M-1 fiber helmet liner during firing practice at Fort Benning, Georgia in 1942. (Courtesy of National Archives, 111-SC-142865)

modification prevented the separation of the liner from the helmet body. Modifications to the M-1 liner included the addition of female snap fasteners to receive the helmet body's web chin strap, inverted khaki, olive drab shade no. 3, cotton webbed A-straps with wire buckles, and a chamois lined leather molded chin cup. The inverted khaki, olive drab shade no. 3, cotton webbed A-straps were secured to the liner by removing four of the suspension washer and rivet assemblies, sandwiching the ends of the A-straps between the suspension washer and rivet assembly, and riveting the assembly together again. The buckles of the webbed strap allowed for adjustment of the chin cup.

While the specifications for the new Parachutist Helmet were not written until June 1942, the first orders for the parachutist helmet were placed in January 1942. When the specification was finally written in June 1942, the designation given to the new Parachutist Helmet was the M-2. The Specification for the M-2 Parachutist Helmet called for the use of fixed D ring chin strap loops in place of the standard loops. These loops would allow the web chin strap to be worn more easily behind the wearer's head. Later, around the spring or summer of 1943, the Airborne Command stated that the standard chin strap loops were satisfactory and the fixed D ring chin strap loops were no longer warranted. In October 1943 production changed to a helmet with hinged chin strap loops. Between January 1942 and December 1944, approximately 148,000 helmet bodies were taken from existing McCord M-1 stocks and modified by McCord to parachutists configuration.

Exterior side view of the parachutist's M-1 fiber helmet liner manufactured by Hawley Products Company and McCord Radiator, ca. January 1942-November 1942.

The modified parachutist helmet liners were also taken from existing stocks. The modified parachutist helmet liners retained the designation M-1. Initially, 43,000 Hawley fiber liners were modified between January 1942 and fall of 1942. The fiber liners were

THE M-1 HELMET

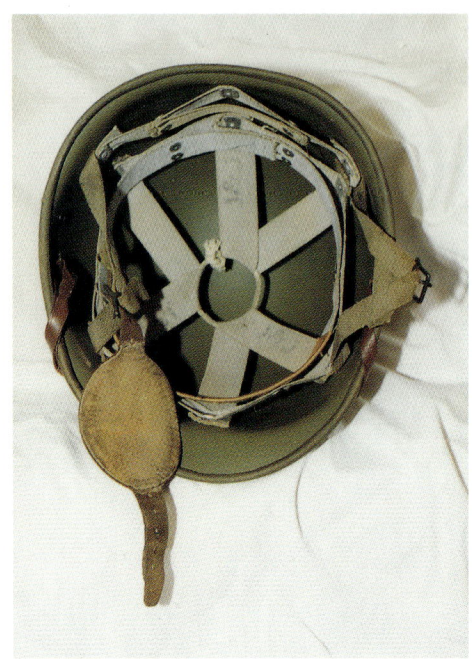

Left: Interior view of the parachutist's M-1 fiber helmet liner, ca. January 1942-November 1942. Center: Interior view of the parachutist M-1 fiber helmet liner, ca. January 1942-November 1942. Note the J&J stamp on the rayon suspension representing the Johnson and Johnson Company. Right: Close-up of the parachutist M-1 fiber helmet liner's cotton, olive drab shade no. 3, inverted A-straps.

modified to the parachutist configuration by the McCord Radiator and Manufacturing Company. The first plastic liners to be modified numbered 75,000 and came from the Inland Manufacturing Division of General Motors. The modification of these plastic liners was also performed by the McCord Radiator and Manufacturing Company during the fall of 1942. From September 1943 until spring of 1944, Westinghouse supplied and modified the last major batch of the parachutist helmet liners. Approximately 30,000 helmet liners were modified by Westinghouse. During the spring of 1944, the chamois lined leather molded chin cup was replaced by a khaki, olive drab shade no. 3, webbed chin strap cup for both production and probable replacement. The change was made due to the cheaper manufacturing cost of the web chin strap cup and approximately 40,000 web chin strap cups were produced during the spring of 1944.

M-1C PARACHUTIST HELMET

In January 1945, the M-1C Helmet was standardized by the U.S. Army. The M-1C was essentially the same as the earlier M-2 Parachutist Helmet and also incorporated all of the other changes found in the standard M-1 helmet. Between January 1945 and August 1945, the U.S. Army procured 392,000 M-1C Helmets.

Slight differences in the M-1C helmets included the use of olive drab shade no. 7 webbing in the manufacture of the liner A-straps, and the use of A-straps with cast buckles.

PARACHUTIST HELMET TOTALS

The wearer of either parachutist helmet, was instructed prior to jumping, to unfasten the strap of the helmet from under the chin and secure the helmet body to the liner by means of the snap fasteners on the short ends of the body's web chin strap. The web liner straps,

Exterior and interior views of the chamois-lined leather chin cup.

CHAPTER SEVEN: PARACHUTIST HELMET

with the chin cup, were to be fastened under the chin, while the helmet body's web chin strap was to be placed behind the head. Both helmets weighed 3 pounds, 2 ounces. The helmets weighed more than their standard counterparts, this being due to the addition of the extra components.

It is probable that an equal number of helmet bodies were produced or modified for every parachutist helmet liner. The total amount of parachutist helmets procured during World War II was approximately 540,000, with more than half being M-1Cs.

Above left: Exterior rear left view of the M-2 helmet body with parachutist M-1 fiber helmet liner. (Courtesy of Ed Hicks)

Left: Interior view of the M-2 helmet body with parachutist M-1 fiber helmet liner. (Courtesy of Ed Hicks)

Above right: Exterior front left view of the M-2 steel helmet body, ca. 1942-1943. This first variation steel helmet body for parachutists has olive drab shade no. 3 webbing, brass fixtures, stainless steel rim, and D-ring chin strap loops. (Courtesy of Michel De Trez)

Close-up of the M-2 helmet body's fixed D-ring loop, web chin strap, and extension with snap fastener, ca. January 1942-summer 1943. (Courtesy of Michel De Trez)

69

THE M-1 HELMET

Paratroopers line up for a practice jump during August 1943. These paratroopers wear the M-2 paratrooper helmet. Note the D-ring on the helmet body and the liner's inverted A-straps with chin cup. (Courtesy of National Archives, 111-C-3613)

CHAPTER SEVEN: PARACHUTIST HELMET

Left: Exterior view of the M-2 helmet body with parachutist M-1 plastic helmet liner. (Courtesy of Ed Hicks) Right: Close-up of the parachutist helmet body's web chin strap and extension with snap fastener, sewn to fixed chin strap loop, ca. summer 1943. Note interior view of Inland Manufacturing Divsion's M-1 parachutist helmet liner, with female snap fastener and cotton, olive drab shade no. 3, A-straps, ca. 1943.

Close-up of the helmet body's fixed loop, web chin strap, and extension with snap fastener, ca. summer 1943-October 1943. (Courtesy of Michel De Trez)

Exterior side view of the parachutist M-1 plastic helmet liner manufactured by the Inland Manufacturing Division of General Motors, ca. 1943. (Courtesy of Michel De Trez)

Left: Close-up of the parachutist helmet liner's wire buckle as used with the M-1 parachutist helmet liner. Note the cotton, olive drab shade no.3, A-strap webbing, and leather chin cup. (Courtesy of Dallas W. Freeborn)

THE M-1 HELMET

Interior view of the parachutist M-1 plastic helmet liner, ca. 1943. This particular liner was manufactured by the Inland Manufacturing Division of General Motors. Components include cotton, olive drab shade no. 3, A-straps with wire style buckles, and chamois-lined leather chin cup. (Courtesy of Dallas W. Freeborn)

Exterior front left view of the parachutist steel helmet body, ca. October 1943-November 1944. This variation steel helmet body for parachutists has olive drab shade no. 3 webbing, brass fixtures, and a stainless steel rim. (Courtesy of Michel De Trez)

Close-up of the helmet body's flexible loop, web chin strap, and extension with snap fastener, ca. October 1943-November 1944. (Courtesy of Michel De Trez)

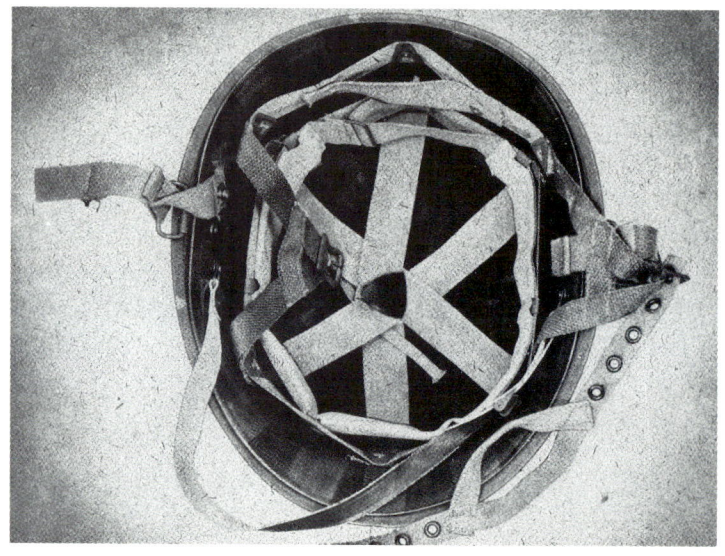

Above: M-1C parachutist helmet. Exterior side view. (Courtesy of U.S. Army Ordnance Department via National Archives, RG 156 Ordnance) Right: M-1C parachutist helmet. Interior view showing fully assembled helmet. (Courtesy of U.S. Army Ordnance Department via National Archives, RG 156 Ordnance)

CHAPTER SEVEN: PARACHUTIST HELMET

Left: Exterior front left view of the M-1C steel helmet body, ca. January 1945-August 1945. This late variation steel helmet body for parachutists has olive drab shade no. 7 webbing, brass fixtures, and a manganese rim with a rear edging butt. Right: Interior side view of the M-1C steel helmet body, ca. January 1945-August 1945.

Close-up of the M-1C helmet body's web chin strap and extension with snap fastener, ca. January 1945-August 1945.

THE M-1 HELMET

Exterior side view of the parachutist plastic helmet liner used with the M-1C and manufactured by Westinghouse, ca. January 1945-August 1945. (Courtesy of Michel De Trez)

Interior side view of the parachutist plastic helmet liner used with the M-1C and manufactured by Westinghouse, ca. January 1945-August 1945. (Courtesy of Michel De Trez)

Above: Close-up of the parachutist helmet liner's cast buckle as used with the M-1C helmet. Note the cotton, olive drab shade no. 7, A-strap webbing, and five grommet open web chin cup. (Courtesy of Dallas W. Freeborn)

Right: Interior view of the parachutist plastic helmet liner used with the M-1C, ca. January 1945-August 1945. This particular liner was manufactured by the Westinghouse Electric Company. Components include cotton, olive drab shade no. 7, A-straps with cast style buckles, and five grommet open web chin cup. (Courtesy of Dallas W. Freeborn)

CHAPTER SEVEN: PARACHUTIST HELMET

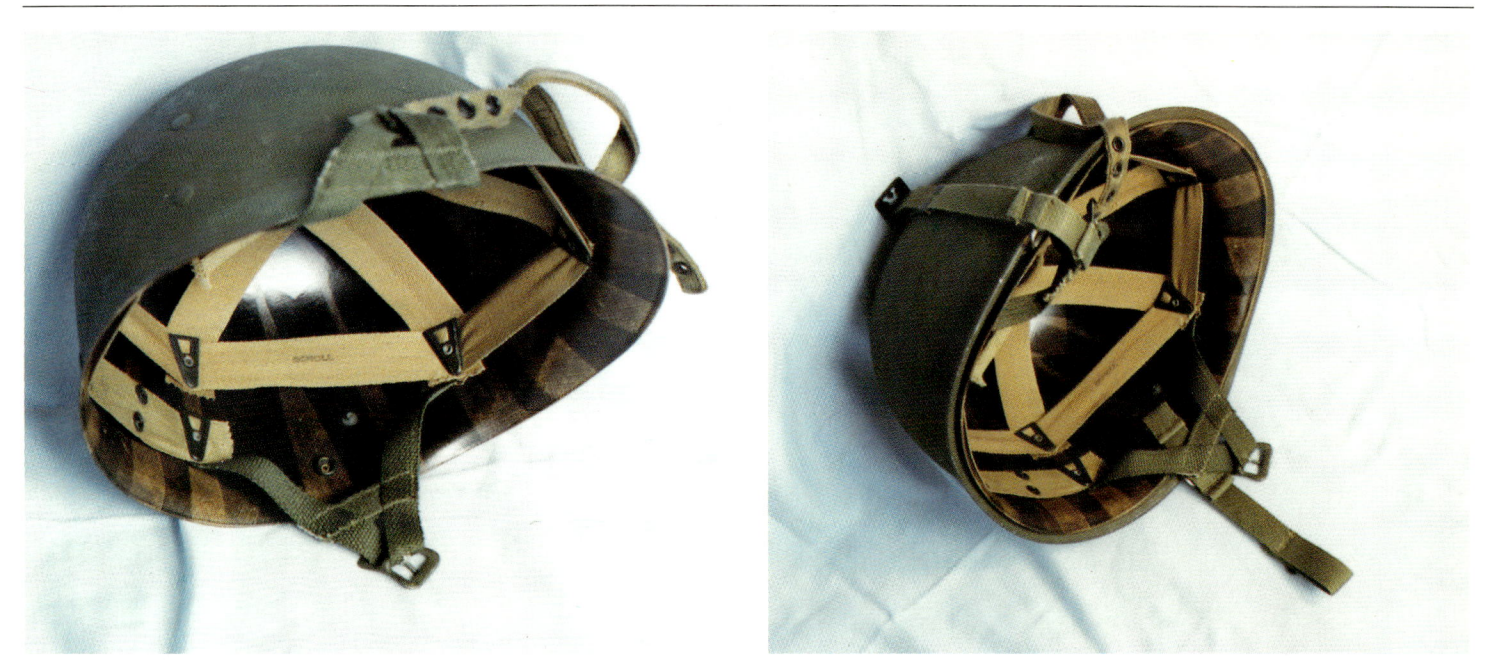

Left: Interior side view of the plastic helmet liner as used with the M-1C showing the parachutist's chin strap unattached and the inverted A-strap webbing secured to suspension A washers and rivets, ca. January 1945-August 1945. (Courtesy of Keith R. Jamieson, M.D.) Right: Interior side view of M-1C helmet body and liner, ca. January 1945-August 1945. (Courtesy of Keith R. Jamieson, M.D.)

An Army paratrooper lines up for a jump during 1945. The paratrooper wears the M-1C paratrooper helmet. Note the open web chin cup. (Courtesy of U.S. Army Signal Corps, 233562, via George A. Petersen)

CHAPTER EIGHT

HELMET CAMOUFLAGE

The completed M-1 Helmet often provided adequate camouflage to the wearer, but by 1944, several services had developed and issued items that provided the helmet with additional camouflage characteristics. The techniques used to enhance the camouflage properties involved the use of camouflage helmet covers, the painting of helmets with a camouflage pattern, and the use of helmet nets.

CAMOUFLAGE HELMET COVERS
In September 1942, the United States Marine Corps adopted the camouflaged helmet cover for use with the M-1 Helmet. The probable use of the helmet cover was to provide the wearer with additional camouflage properties in at least two different types of environments. The cover was manufactured in two different color combinations, both using the same camouflage pattern. The colors of the helmet cover suggest that the cover was to be used in either a jungle type environment or a beach/desert environment.

The cover was manufactured of cotton Herringbone twill cloth, and consisted of a "greenside" and a "brownside." The "greenside" consisted of a light green background with a four color pattern of medium and light, brown and green. The "brownside" consisted of a tan background with a three color pattern of medium, light, and very light brown. The cover was manufactured of two identical halves, which when sewn together created a dome-like piece of material that fitted neatly over the helmet body. The cover included flaps which were sandwiched between the helmet liner and the helmet body when the cover was used, or which could be left out to cover the neck or ears if the wearer chose to break up the outline of the helmet. The cover was produced with and without buttonholes. These buttonholes provided a means of fastening additional camouflage material, such as surrounding vegetation. The cover contained sixteen crown buttonholes and one buttonhole for each flap. The helmet cover also appeared with and without markings. The markings were applied after manufacture. There were at least two types of Marine Corps markings stamped on the helmet cover in two locations, one on each side. The most common marking used consisted of the basic "eagle, globe, and anchor" with no riband or rope, and measured approximately 1.75" by 1.75". The other mark-

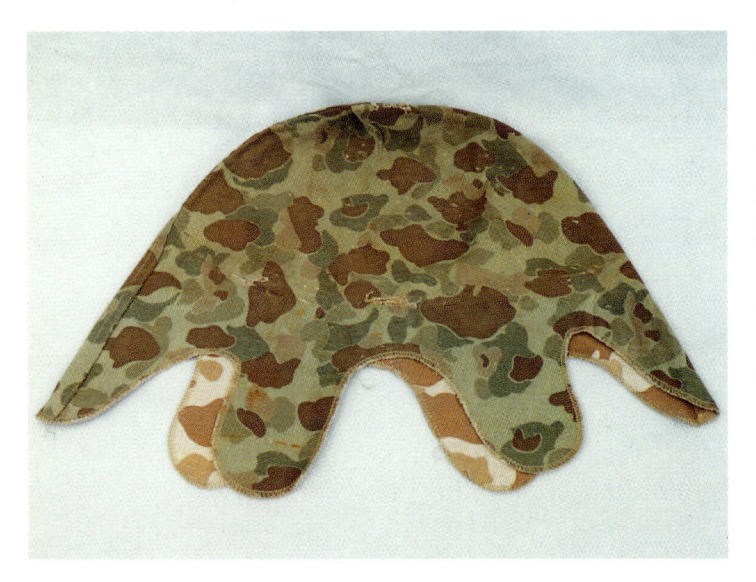

U.S. Marine Corps camouflage helmet cover, green side out, September 1942-August 1945.

Front left view of M-1 helmet with U.S. Marine Corps camouflage helmet cover, brown side out, September 1942-August 1945.

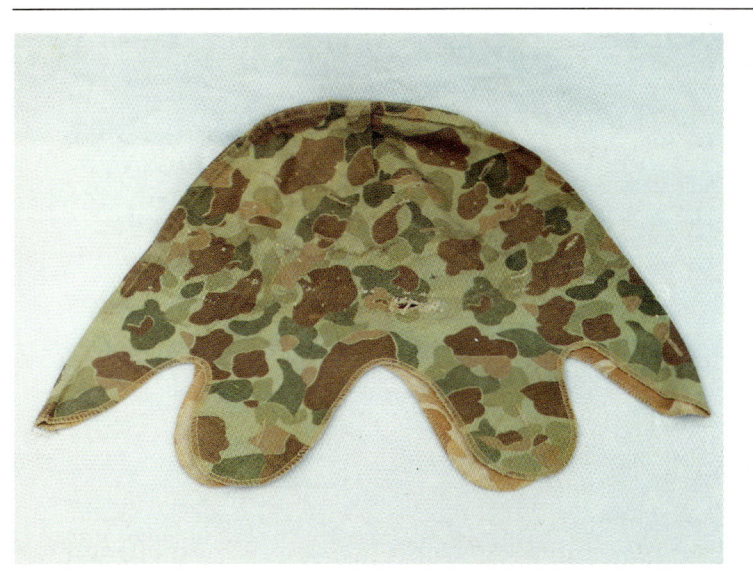

CHAPTER EIGHT: HELMET CAMOUFLAGE

U.S. Marine Corps camouflage helmet cover with buttonholes, green side out, September 1942-August 1945.

Front left view of M-1 helmet with U.S. Marine Corps camouflage helmet cover with buttonholes, green side out, September 1942-August 1945.

ing used appeared the same as the above marking, except being considerably larger. Other than the Marine Corps symbol, the helmet cover appeared with no other markings.

A second type of helmet cover was produced for the Marine Corps for the M-1 helmet and was known as the mosquito hat net. The mosquito hat net helmet cover utilized the same camouflage pattern as the other Marine Corps helmet covers, but consisted of two major parts, a cotton printed helmet cover and a fine mesh skirt. The cotton print was non-reversible, having only the "greenside" printed on its quarters. Sewn to this outer "greenside", around its entire circumference was a green looped foliage band. The two piece mesh skirt, believed to be made of a nylon material, was also printed with only the "greenside" pattern. Attached to the bottom of the mesh skirt were green cotton laces, provided so that the wearer could secure the mesh to his upper body, so as to prevent the mesh from rising.

The Marine Corps camouflage helmet covers were manufactured between 1942 and 1945. The actual number produced has not

A Marine Corps Dog Handler wears the Marine camouflage helmet cover over his M-1 helmet. His cover is the version that contains no buttonholes, and he wears the brownside out. (Courtesy of U.S. Marine Corps)

Marine Corporal Fenwick H. Dunn wears the Marine Corps camouflage helmet cover with buttonholes and the greenside out. (Courtesy of U.S. Marine Corps)

THE M-1 HELMET

been found, nor the names of the companies that may have produced them, but it is known that the cover was only issued to Marine Corps troops during World War II.

CAMOUFLAGE HELMET LINER

As early as 1942, the U.S. Army had sought some additional camouflage properties for the helmet, and by late February 1942, procurement was started on M-1 helmets with a special painted camouflage pattern. Between February 1942 and March 1943, 854,225 helmet liners were camouflaged and probably an equal number of helmet bodies were also camouflaged. The helmet bodies were likely selected from existing stock. The liners were completed Westinghouse models that underwent application of a spray painted template camouflage pattern at Westinghouse. In March 1944, procurement of the camouflaged helmet liner and helmet bodies was discontinued. Also in March 1944, 300,000 of the camouflaged helmet liners were repainted once again with olive drab, as the Quartermaster reported that use of netting provided a more effective camouflage. The helmet was often also referred to as the M-1 Jungle helmet liner.

CAMOUFLAGE HELMET NETTING

By the spring of 1944, the U.S. Army had been effectively using British helmet netting to upgrade the camouflage properties of the M-1 Helmet. The British manufactured helmet netting was khaki in color, and used a open mesh woven with 0.75" openings. In the summer of 1944, the U.S. Army Quartermaster Board had set out to test a helmet net which they had developed for the M-1 Helmet. The tests were conducted at Camp Indian Bay, Florida, and the results were in by September 1944. The results showed that the helmet net had performed satisfactorily, and that the net was to be recommended for issue. The Army issued this helmet net to their troops in the fall of 1944. The net was to be used in any operation where camouflage was beneficial.

Major General G.B. Erskine, at left and Lieutenant General Holland Smith, at right on Iwo Jima, February 1945. Major General Erskine wears a Marine Corps camouflage helmet cover with buttonholes and the greenside out, while Lieutenant General Smith wears a Hawley tropical fiber hat with Marine Corps insignia. (Courtesy of U.S. Marine Corps)

The Camouflage Helmet Net, identified by the Quartermaster as Item No. 234, was of "Cotton fabric, open mesh weave, O.D. shade no. 7, with cotton elastic band, O.D. shade no. 7, attached 8" from rear of net. Manufactured in 1 size, hexagonal in shape, 28" in diameter, unfinished edges, no seams. Constructed with cutouts on sides 6" deep by 2" wide, to permit adjustment around helmet strap." The Camouflage Helmet Net was issued with three components, and they included the net, the attached cotton elastic band, and an instruction tag that advised the wearer on its use.

The Camouflage Helmet Net was found to provide adequate camouflage when used with vegetation. The wearer was instructed to sandwich the ends of the net between the helmet body and the liner. The wearer could also allow a portion of the netting to hang

U.S. Marine Corps camouflage helmet cover with Eagle, Globe and Anchor stamp applied, brown side out, September 1942-August 1945.

U.S. Marine Corps camouflage helmet cover with Eagle, Globe and Anchor stamp applied, green side out, September 1942-August 1945.

CHAPTER EIGHT: HELMET CAMOUFLAGE

View of two U.S. Marine Corps camouflage helmet covers with different sized Eagle, Globe and Anchor stamps applied.

Front view of M-1 helmet with U.S. Marine Corps camouflage helmet cover with large Eagle, Globe and Anchor stamp applied, green side out, September 1942-August 1945.

loosely over the back of the neck. When worn in this fashion, the net not only broke up the outline of the head and neck, but the net also afforded protection to the neck itself against insects and possibly the sun. While the net did tend to fray or snag objects, it also reduced the noise of the helmet when it was struck by an object. The helmet net also reduced the glare of the helmet. The net could be worn over the elastic band, or the other way around.

The U.S. Army Camouflage Helmet Net was manufactured between September 1944 and summer of 1945. The actual number produced has not been found, nor the names of the companies that may have produced the Camouflage Helmet Net. The net appeared to have been issued exclusively to U.S. Army troops during World War II.

Bottom left: Front view of M-1 helmet with U.S. Marine Corps camouflage helmet cover with small Eagle, Globe and Anchor stamp applied, brown side out, September 1942-August 1945. Bottom right: Front left view of M-1 helmet with U.S. Marine Corps camouflage helmet cover with small Eagle, Globe and Anchor stamp applied, brown side out, September 1942-August 1945.

Front right view of M-1 helmet with U.S. Marine Corps camouflage helmet cover with large Eagle, Globe and Anchor stamp applied, green side out, September 1942-August 1945.

THE M-1 HELMET

CHAPTER EIGHT: HELMET CAMOUFLAGE

THE M-1 HELMET

Previous: Soldiers of the 2nd Marine Division on Tarawa during November 1943. Note the Marine Corps mosquito hat net camouflage helmet cover worn by the Marine standing at the far left. (Courtesy of National Archives, 127-N-63788)

U.S. Marine Corps mosquito hat net camouflage helmet cover, ca. 1943-1945.

M-1 helmet body with U.S. Marine Corps mosquito hat net camouflage helmet cover, ca. 1943-1945.

The 1st Marine Division Command displaying various helmet liners and camouflage. From left to right, Lieutenant Colonel Conoley wearing a fiber liner, Lieutenant Colonel Puller wearing a Westinghouse Camouflage plastic helmet liner, unidentified officer wearing a fiber liner, unidentied officer wearing a Westinghouse Camouflage plastic helmet liner, unidentifed officer wearing the complete helmet with the Marine Corps Camouflage helmet cover, and Captain Buckley wearing a fiber liner. (Courtesy of National Archives, 127-GW-946-72616)

CHAPTER EIGHT: HELMET CAMOUFLAGE

Camouflaged plastic helmet liner with spray painted stenciled pattern worn by USMC Navajo code talkers. (Courtesy of U.S. Marine Corps)

Exterior front left view of the M-1 plastic helmet liner with stenciled camouflage pattern manufactured by Westinghouse Electric Company. (Courtesy of Dallas W. Freeborn)

Exterior right side view of the Westinghouse M-1 plastic camouflage helmet liner. (Courtesy of Keith R. Jamieson, M.D.)

Exterior front view of the Westinghouse M-1 plastic camouflage helmet liner (Courtesy of Keith R. Jamieson, M.D.)

THE M-1 HELMET

Exterior right top view of the Westinghouse M-1 plastic camouflage helmet liner. (Courtesy of Dallas W. Freeborn)

Exterior rear left view of the Westinghouse M-1 plastic camouflage helmet liner. (Courtesy of Keith R. Jamieson, M.D.)

Exterior top view of the Westinghouse M-1 plastic camouflage helmet liner. (Courtesy of Keith R. Jamieson, M.D.)

CHAPTER EIGHT: HELMET CAMOUFLAGE

CAMOUFLAGE HELMET BAND

The Camouflage Helmet Band, which was used with the Camouflaged Helmet Net, was developed and issued prior to the development of the net. The band was a standard item as early as fall of 1942. The band was elastic and constructed of Neoprene, a synthetic rubber. The elastic strip was formed into a band by joining the ends with a box tack stitch with diagonals. The finished band was 23" in circumference and was green in color, olive drab shade no. 7. The band was to go over the helmet body or the liner, and to hold branches, leaves or other vegetation for camouflage.

The U.S. Army Camouflage Helmet Band was manufactured between the fall of 1942, and the summer of 1945. Like the net, the actual number of bands produced has not been found, nor the names of the companies that may have produced the Camouflage Helmet Band. The band also appeared to have been issued only to U.S. Army troops during World War II.

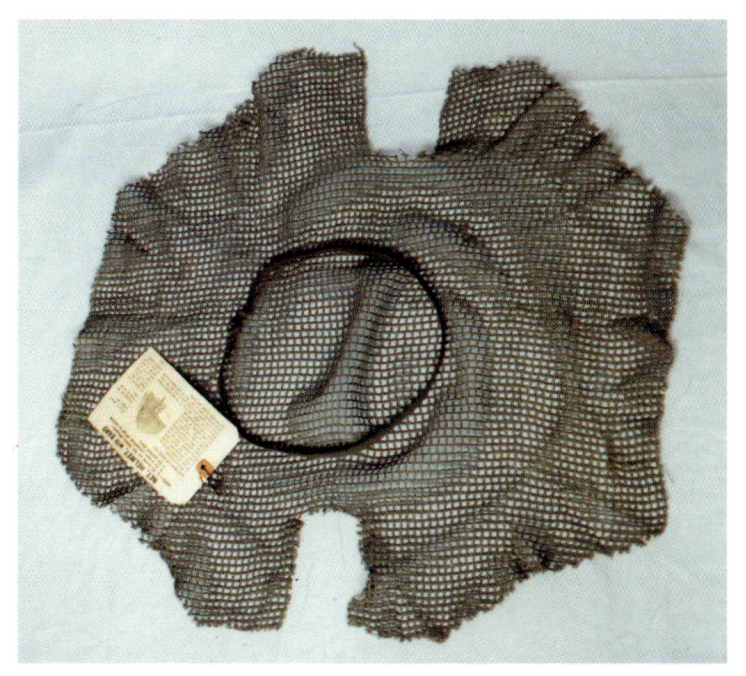

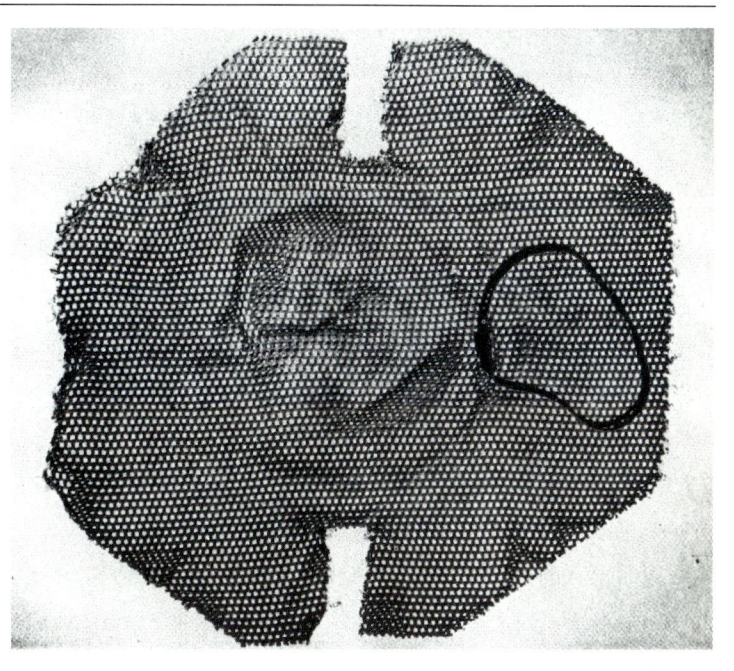

Camouflage helmet net with attached camouflage helmet band. (Courtesy of U.S. Army Quartermaster Corps, Library Collections of the U.S. Army Quartermaster Museum, Fort Lee, Virginia)

Item 234, camouflage helmet net, band and instruction tag. (Courtesy of Keith R. Jamieson, M.D.)

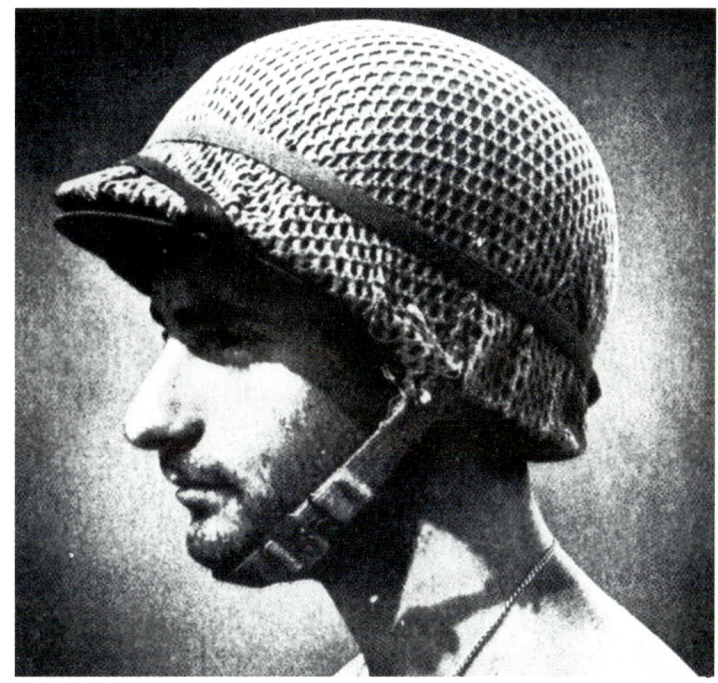

Camouflage helmet net shown on the M-1 helmet with all netting ends sandwiched between the helmet body and helmet liner, and camouflage helmet band worn over the net. (Courtesy of U.S. Army Quartermaster Corps, Library Collections of the U.S. Army Quartermaster Museum, Fort Lee, Virginia)

THE M-1 HELMET

Exterior rear right view of M-1 helmet with camouflage helmet net, band and instruction tag, Item 234. (Courtesy of Keith R. Jamieson, M.D.)

Exterior front left view of M-1 helmet with camouflage helmet net and band, Item 234. (Courtesy of Keith R. Jamieson, M.D.)

M-1 helmet with camouflage helmet net, Item 234, without helmet Band, ca. September 1944-August 1945.

Camouflage helmet net shown on the M-1 helmet with all netting ends sandwiched between the helmet body and helmet liner, and camouflage helmet band worn underneath the net. (Courtesy of U.S. Army Quartermaster Corps, Library Collections of the U.S. Army Quartermaster Museum, Fort Lee, Virginia)

Interior view of the M-1 helmet with camouflage helmet net, Item 234, without helmet Band. This view shows how the net was sandwiched between the steel body and the liner.

86

CHAPTER EIGHT: HELMET CAMOUFLAGE

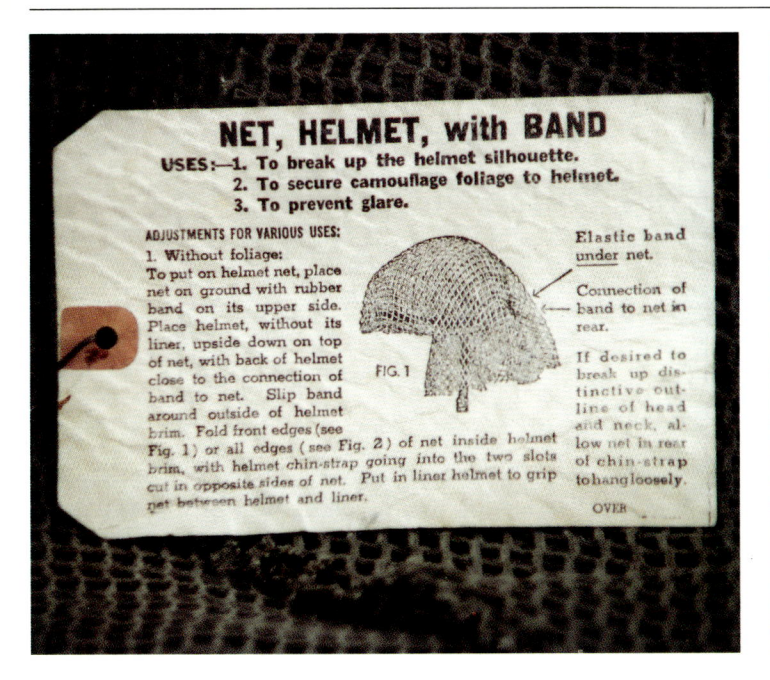

Camouflage helmet net, Item 234, instruction tag, front view.

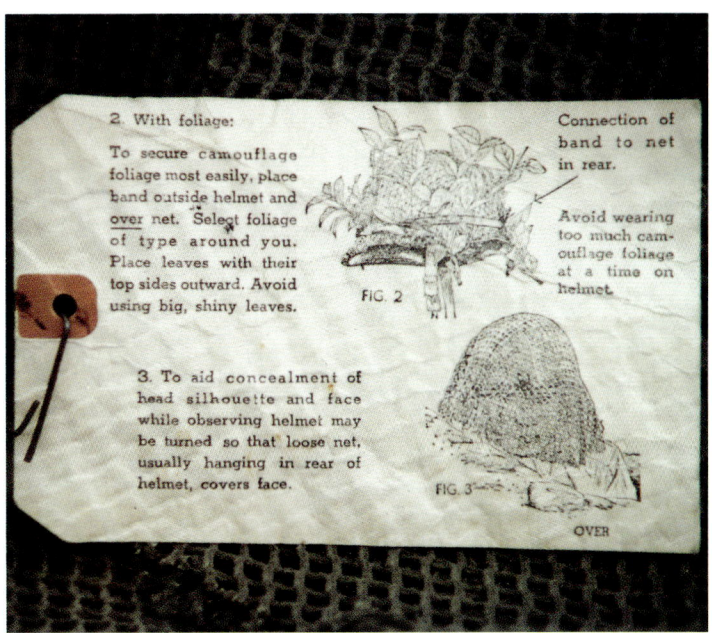

Camouflage helmet net, Item 234, instruction tag, rear view.

Camouflage helmet net worn over the band with the front edge tucked in between the helmet body and liner and the back edge hanging loosely over the neck and shoulders. (Courtesy of U.S. Army Quartermaster Corps, Library Collections of the U.S. Army Quartermaster Museum, Fort Lee, Virginia)

Camouflage helmet net from the S-6283 information series. Note that this was the picture that was used to illustrate the instruction tag. (Courtesy of War Department Office of War Information)

87

THE M-1 HELMET

Left: A G.I. receiving new uniform items, including the camouflage helmet net with band during the winter 1944-1945. (Courtesy of U.S. Army) Right: A soldier of the U.S. Army 96th Infantry Division wears the camouflage helmet band over his M-1 helmet during fighting on Okinawa in May 1945. (Courtesy of U.S Army Signal Corps)

Two Marines cooking hot chow wear their Marine Corps camouflage helmet covers. Marine on the left wear a camouflage helmet band over his helmet cover. (Courtesy of U.S. Marine Corps)

CHAPTER EIGHT: HELMET CAMOUFLAGE

Camouflage helmet net worn over the M-1 helmet in combat. Note the helmet body's chin strap is worn under the helmet net. (Courtesy of U.S. Army Signal Corps, 189265, via George A. Petersen)

CHAPTER NINE
FIELD MODIFICATIONS AND FIELD MARKINGS

When the M-1 helmet was used by combat troops, it often received additional field modifications and field markings. This chapter presents a sample of some of the most commonly applied modifications and markings applied to the M-1 helmet.

The United States Army Air Force issued the M-1 helmet to all of the aircrews. Shown here are two members of a B-17 aircrew with experimental body armor and the standard M-1 helmet. (Courtesy of U.S. Army Air Force)

CHAPTER NINE: FIELD MODIFICATIONS AND FIELD MARKINGS

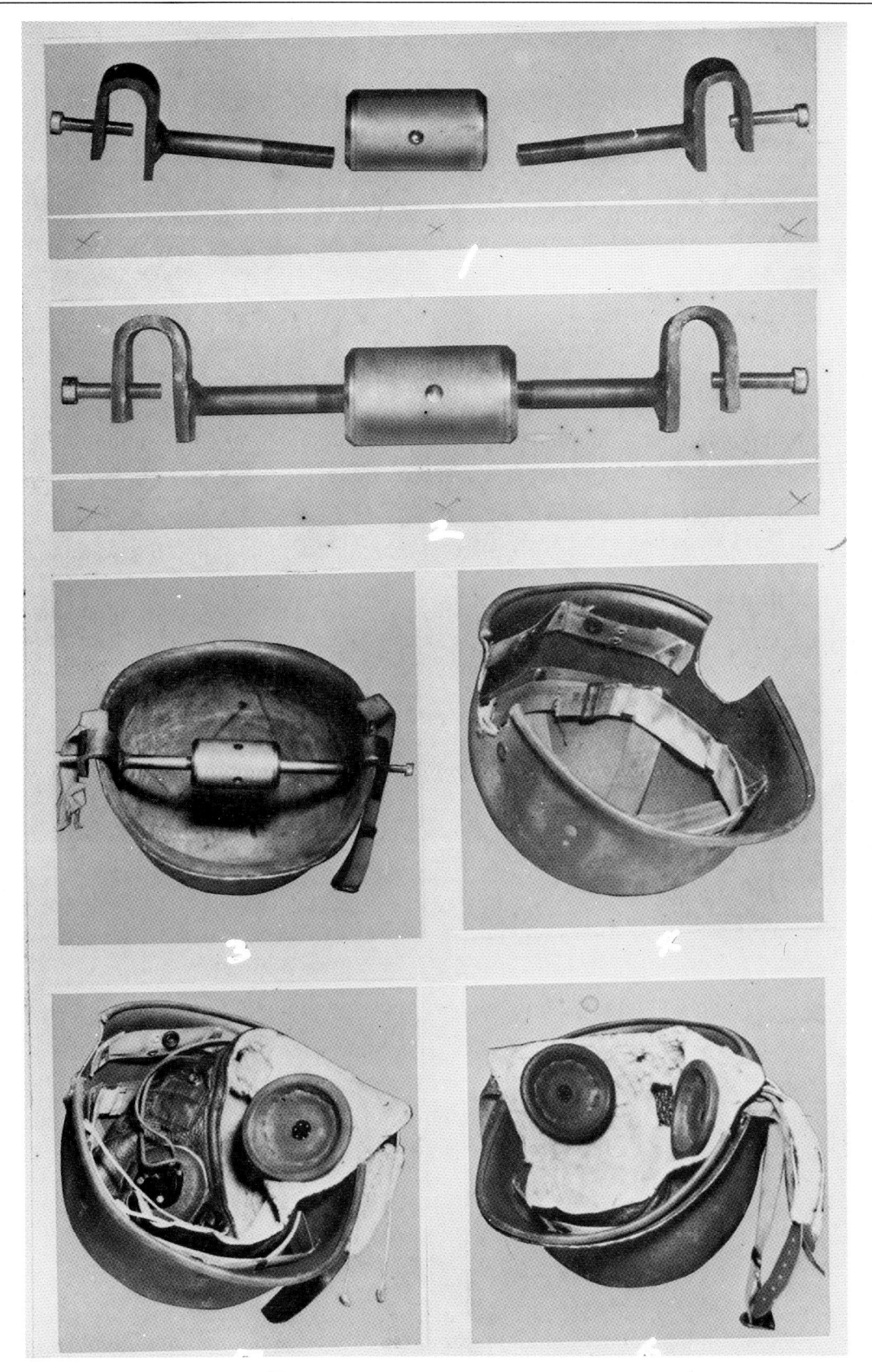

USAAF aircrews began to modify standard M-1 helmets during 1943, due to the fact that an unmodified M-1 helmet caused severe discomfort and pressure to the airmen when worn with headgear. The M-1 helmet was modified by first having its sides spread through use of a screw-jack. Next sections were cut from the sides of the liner, in this photo a Hawley fiber liner is shown. Finally the liner head band could be adjusted to accommodate a flying helmet. (U.S. Air Force Photo Collection, USAF Neg. No. 8316AC, courtesy of National Air and Space Museum, Smithsonian Institution)

91

THE M-1 HELMET

Exterior side view of the USAAF modified plastic helmet liner. The liner's sides were removed to permit aircrews to wear headgear without discomfort. (Courtesy of Jon A. Maguire)

Interior view of the USAAF modified plastic helmet liner. This view clearly identifies this plastic helmet liner as an early St. Clair low-pressure liner. (Courtesy of Jon A. Maguire)

Exterior front view of the USAAF modified M-1 helmet body. (Courtesy of Jon A. Maguire)

Exterior front right view of the USAAF modified M-1 helmet body. (Courtesy of Jon A. Maguire)

CHAPTER NINE: FIELD MODIFICATIONS AND FIELD MARKINGS

Exterior rear right view of the USAAF modified M-1 helmet body. (Courtesy of Jon A. Maguire)

Interior view of USAAF modified M-1 helmet complete. (Courtesy of Jon A. Maguire)

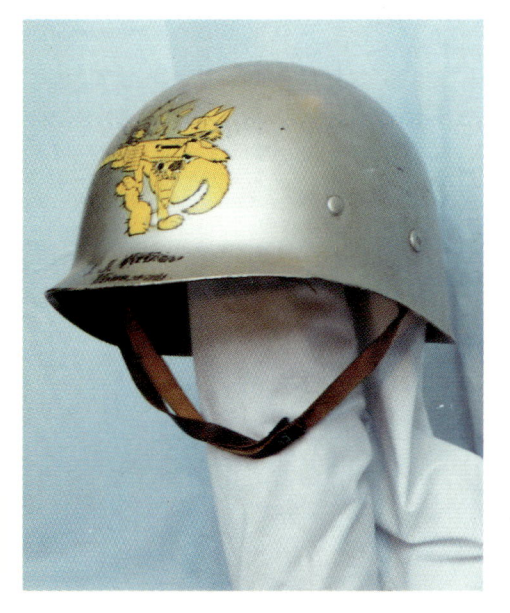

Exterior front left view of an AAF Gunnery School M-1 helmet liner with silver finish and yellow decal. (Courtesy of Keith R. Jamieson, M.D.)

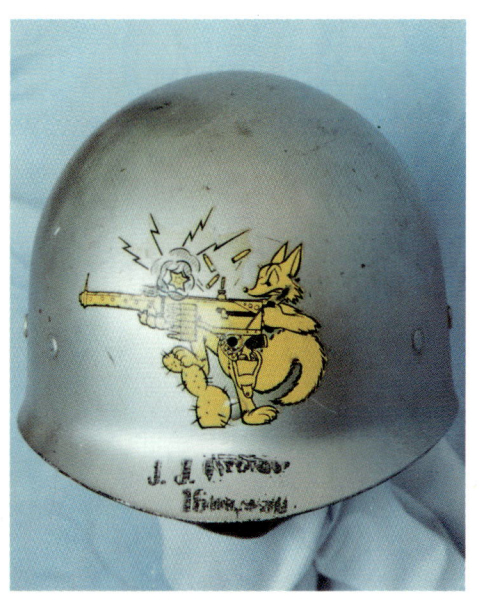

Exterior front view of an AAF Gunnery School M-1 helmet liner with silver finish and yellow decal. (Courtesy of Keith R. Jamieson, M.D.)

M-1 helmet liner from Kingman Army Air Field Gunnery School, ca. 1942-1945. (Courtesy of Mohave Museum of History & Arts - Kingman, AZ)

93

THE M-1 HELMET

U.S. Army soldiers standing in a chow line during training exercises. The soldiers wear M-1 plastic helmet liners, which are covered by various U.S. Army applied identification numbers. (Courtesy of National Archives, 111-C-268)

CHAPTER NINE: FIELD MODIFICATIONS AND FIELD MARKINGS

U.S. Army applied identification number used on the M-1 helmet body.

U.S. Army applied identification numbers on the M-1 helmet body, ca. 1941-1943.

U.S. Army applied identification numbers on the M-1 plastic helmet liner, ca. fall 1942. This particular early plastic liner was manufactured by the Inland Division of General Motors, and has the early dark olive drab paint without the resin texture.

THE M-1 HELMET

Exterior front left view of standard Military Police markings on an M-1 helmet. (Courtesy of Ed Hicks)

Exterior rear right view of standard Military Police markings on an M-1 helmet. (Courtesy of Ed Hicks)

U.S. Military Policeman wearing an M-1 helmet with M.P. Markings. (Courtesy of U.S. Army)

CHAPTER NINE: FIELD MODIFICATIONS AND FIELD MARKINGS

8th Infantry Division MP helmet, ca. 1944-1945. This helmet has the white MP stripe around the helmet and unit shield on both sides. (Courtesy of Ed Hicks)

U.S. Army Military Police plastic helmet liner, ca. 1944 -1945.

M-1 helmet body and liner painted overall white with black MP markings. (Courtesy of Michel De Trez)

97

THE M-1 HELMET

U.S. Army Medical Corpsman tending to the wound on an injured soldier. Note M-1 helmet with International Red Cross markings. (Courtesy of U.S. Army)

International Red Cross markings painted on Medic's M-1 helmet. (Courtesy of Michel De Trez)

Painted Medic's insignia, three white ovals with a red cross, from the 551st Parachutist Infantry Battalion, 1st Airborne Task Force, Southern France. (Courtesy of Michel De Trez)

Medical Corpsman helmet with International Red Cross markings, on late model M-1 helmet, ca. 1944-1945. (Courtesy of Ed Hicks)

CHAPTER NINE: FIELD MODIFICATIONS AND FIELD MARKINGS

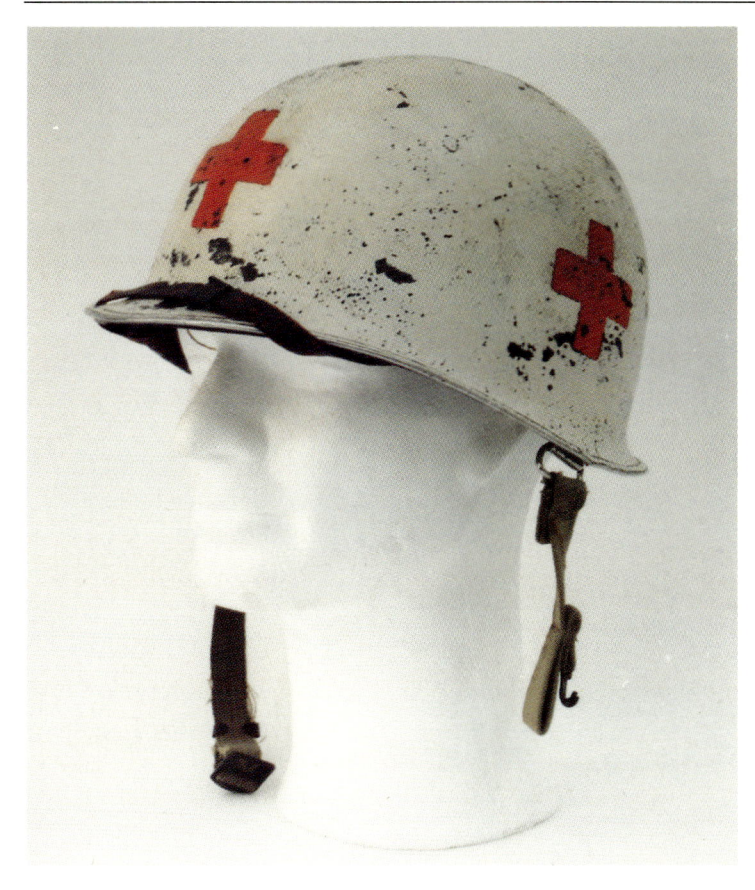

M-1 helmet body painted overall white with red cross markings. (Courtesy of Michel De Trez)

M-1 helmet liner painted overall white with red cross markings. (Courtesy of Michel De Trez)

Front right view M-1 plastic helmet liner with Civil Defense markings and chin strap extended.

Front view M-1 plastic helmet liner with Civil Defense markings, ca. 1944-1945. This particular helmet represents the Auxiliary Fire Department Civil Defense Unit from Huntington Park, California.

99

THE M-1 HELMET

2nd Ranger Battalion insignia painted on rear of M-1 helmet, ca. June 1944. (Courtesy of Ed Hicks)

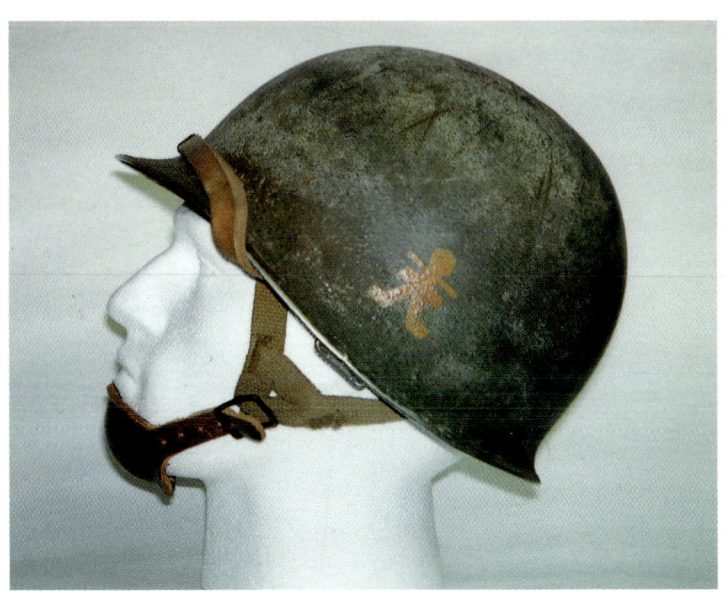

509 Parachute Infantry Regiment "Gingerbreadman" insignia on parachutist helmet. (Courtesy of Gary Howard)

Exterior front right view of M-1 helmet body with 3rd Infantry Division and Lieutenant's markings. (Courtesy of Keith R. Jamieson, M.D.)

A soldier of the 3rd Infantry Division wears painted divisional insignia on his M-1 helmet. (Courtesy of U.S Army Signal Corps)

CHAPTER NINE: FIELD MODIFICATIONS AND FIELD MARKINGS

Interior side view of M-1 helmet body with 3rd Infantry Division and Lieutenant's markings. (Courtesy of Keith R. Jamieson, M.D.)

M-1 helmet with painted Captain's bars insignia, ca. 1941-1943.

M-1 helmet body with painted Lieutenant Colonel markings.

A U.S. Army soldier wearing an M-1 helmet with painted vertical bar insignia in the European Theater of Operations indicating that he was an officer. (Courtesy of U.S. Army)

Rear view of Lieutenant Colonel's M-1 helmet with vertical white bar used to identify officers in the European Theater.

101

THE M-1 HELMET

U.S Army nurses holding the M-1 helmet. Note the painted officer insignia on the front and rear of the helmets. (Courtesy of U.S. Army Signal Corps, 191711, via George A. Petersen)

Opposite: General George S. Patton wearing his highly decorated M-1 helmet liner. (Courtesy of National Archives, 111-C-3538)

CHAPTER NINE: FIELD MODIFICATIONS AND FIELD MARKINGS

THE M-1 HELMET

Above left: Welded Warrant Officer's bar on an M-1 helmet. (Courtesy of Michel De Trez) Above right: Welded First Lieutenant's bar on an M-1 helmet. (Courtesy of Michel De Trez) Right: Welded Captain's bars on an M-1 helmet. (Courtesy of Michel De Trez)

Below: Welded and painted insignia on an M-1 helmet body and M-1 fiber helmet liner worn by Major Hartford T. Salee, 327th Glider Infantry Regiment, 101st Airborne. (Courtesy of Michel De Trez)

CHAPTER NINE: FIELD MODIFICATIONS AND FIELD MARKINGS

Welded and painted insignia on an M-2 helmet body and parachutists fiber liner worn by Colonel George Van Horn Moseley Jr., commanding officer of the 502 PIR, 101st Airborne. (Courtesy of Michel De Trez)

Welded and painted insignia on an M-1 helmet body and M-1 fiber helmet liner worn by Colonel L. A. Hawkings. (Courtesy of Michel De Trez)

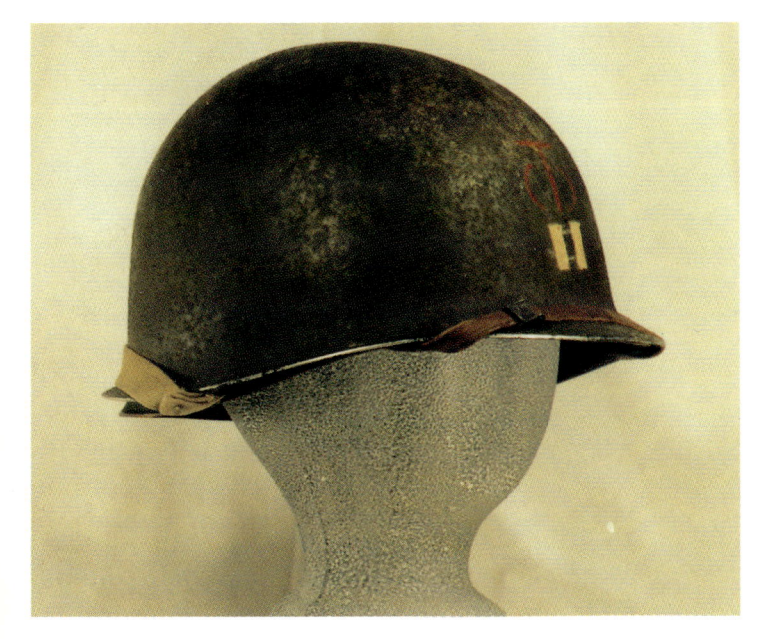

Medical cloth tape Captain's bars and painted 90th Division markings on an M-1 helmet. (Courtesy of Chris Armold)

Medical cloth tape vertical bar on an M-1 helmet used to identify officers in the European Theater. (Courtesy of Chris Armold)

105

THE M-1 HELMET

Aviator Alexander Vraciu indicates his score while deck crew members around him on the USS Lexington wear the M-1 helmet. (Courtesy of U.S. Navy Department via National Archives)

U.S. Navy anti-aircraft gunners were issued M-1 helmets in addition to the MkII Talker helmet. The MkII helmet was also manufactured by McCord. (Courtesy of U.S. Navy Department via National Archives)

Left: U.S. Navy M-1 helmet, ca. 1944-1945. Note that the shell is a late variation made with a manganese rim with rear edging butt, and olive drab shade no. 3 webbing. This particular helmet was manufactured sometime around winter 1944-1945. Right: Right front view of a U.S. Navy M-1 helmet, ca. 1944-1945. Note the bullet hole near the crown.

CHAPTER TEN

TOY AND WORK HELMETS

As manufacturing of the plastic helmet liners got underway, initial production of the plastic helmet liner suffered from high imperfection rate. These imperfect helmets could be salvaged for removable parts, such as suspension, and neck strap, and then either burned or modified of for other purposes. With permission from the Chicago Quartermaster Depot, the plastic helmet liner manufacturers could dispose of these imperfect helmet liners by altering their shape and utilizing unacceptable parts for the suspension, then putting them on the civilian market, and selling them as children's toy helmets or work helmets. Altering their shape usually meant cutting down the sides and rear of the plastic helmet liner. As a result there were a large number of helmets with varying configurations which were sold as toy or work helmets.

Exterior and interior views of a St. Clair toy or work M-1 plastic helmet liner, ca. April 1942.

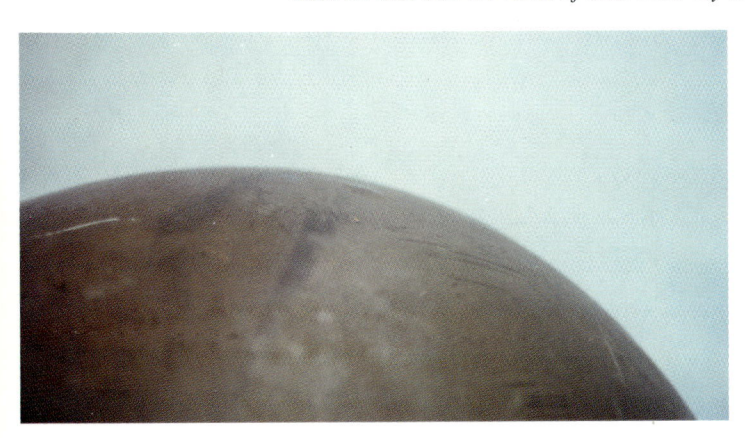

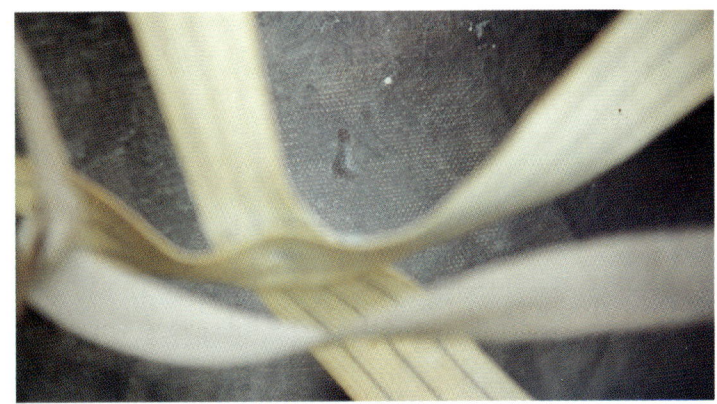

Exterior and interior views of blemish found on the crown of a St. Clair plastic helmet liner.

THE M-1 HELMET

Interior front view of a toy or work St. Clair M-1 plastic helmet liner, with apple green exterior paint and yellow eagle insignia. (Courtesy of Keith R. Jamieson, M.D.)

Exterior front left view of a toy or work St. Clair M-1 plastic helmet liner, with apple green exterior paint and yellow eagle insignia. (Courtesy of Keith R. Jamieson, M.D.)

Exterior front view of a toy or work St. Clair M-1 plastic helmet liner, with apple green exterior paint and yellow eagle insignia. (Courtesy of Keith R. Jamieson, M.D.)

Interior side view of a toy or work St. Clair M-1 plastic helmet liner, with olive drab exterior paint. (Courtesy of Keith R. Jamieson, M.D.)

Left: Interior side view of a toy or work St. Clair M-1 plastic helmet liner, with buff color paint inside and out. This toy helmet utilizes the standard perpendicular-sided washers to secure its' oilcloth suspension. (Courtesy of Keith R. Jamieson, M.D.) Center: Exterior front right of a toy or work St. Clair M-1 plastic helmet liner, with buff color paint. (Courtesy of Keith R. Jamieson, M.D.) Right: Interior side view of a toy or work St. Clair M-1 plastic helmet liner, with olive drab paint on the outside. (Courtesy of Keith R. Jamieson, M.D.)

108

CHAPTER ELEVEN

HELMET BODY AND LINER PRODUCTION: 1951-1952

In March 1951, production of the M-1 helmet resumed. This Korean War era production version of the M-1 helmet incorporated all of the changes and recommendations brought forth from experience gained with the World War II M-1 helmet.

This new production version of the M-1 helmet resembled its World War II forebears, but differed in production items and features. The helmet body's exterior paint was mixed with sand aggregate as opposed to cork, which was used during World War II production. The helmet body's exterior paint was also slightly lighter in color than the World War II production helmet body. The Korean War era production helmet body color was olive drab color no. 319. The helmet body also incorporated the T1 chin strap release and T1 chin strap fastener as production items. The items were previously non-production items, accessories and components developed during the later part of World War II, and did not see much field use.

The production chin strap fastener was made of steel painted a light olive drab, as were chin strap buckle and flat end clip.

The Korean War production liner also looked similar to the final production World War II plastic liner, but differed by incorporating several additional changes. All of the liners interior webbing, the suspension, neck strap, head band and neck band, were manufactured using olive drab shade no. 7 cotton single Herringbone Twill. The leather liner chin strap also had manufacturing changes. The liner chin strap buckle and the chin strap holders were manufactured of steel and painted black. Finally the Korean War era production liners contained two types of manufactures' stamps. The manufacturing stamp appeared as a molded stamp on the interior crown and was often accompanied by an ink stamp on a square piece of impregnated material attached to the interior of the plastic helmet liner.

Left: M-1 helmet complete, ca. 1952. Clearly visible are all the accessory fixtures developed during World War II and now put into production on the Korean War era M-1 helmet. Note that the exterior paint on the body is mixed with a sand aggregate to produce a smaller textured finish. Right: Interior view of the M-1 helmet complete, ca. 1952.

THE M-1 HELMET

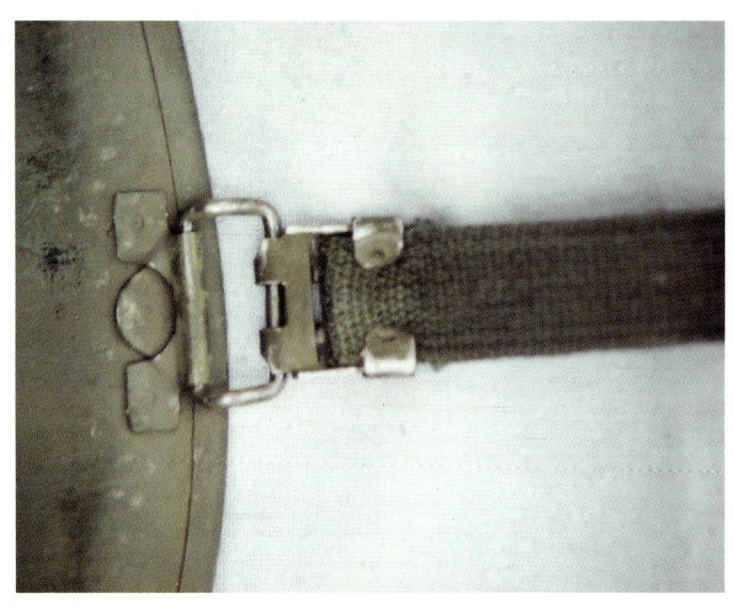

Left: Interior view of the M-1 helmet body, ca. 1951-1952. This view shows the cotton chin strap webbing in olive drab shade no. 7, with steel chin strap fasteners attached. All of the fixtures on the body were made of steel and painted a light olive drab. Right: M-1 helmet body chin strap fastener and flexible loop, ca. 1951-1952. The chin strap fastener was made of steel painted light olive drab.

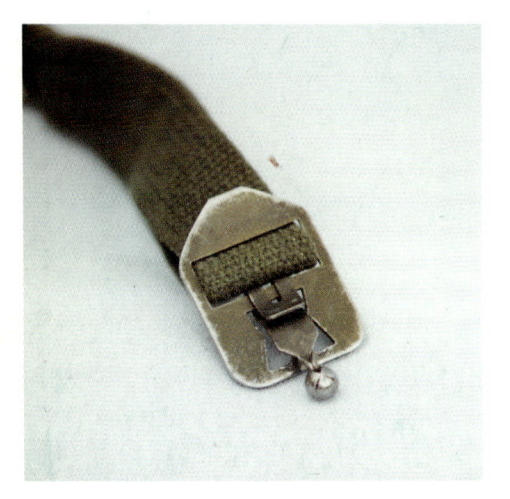

M-1 helmet body chin strap buckle with ball tongue, ca. 1951-1952. The chin strap buckle was stamped and made of steel painted light olive drab.

Exterior view of the M-1 plastic helmet liner, 1952. The liner is painted olive drab with resin texture.

Interior view of the M-1 plastic helmet liner, 1952. This particular liner was manufactured by Westinghouse, and consisted of cotton single Herringbone Twill webbing in olive drab shade no. 7.

110

CHAPTER ELEVEN: HELMET BODY AND LINER PRODUCTION: 1951-1952

M-1 helmet liner head band, ca. 1952. The head band consisted of a bar buckle made of brass painted olive drab, spring clips made of steel painted olive drab, and cotton single Herringbone Twill in olive drab shade no. 7.

M-1 helmet liner neck band, ca. 1952. The head band consists of a bar buckle made of brass painted olive drab and brass snap fasteners coated with a mildew inhibitor. The stamp marking contains the newly standardized MIL SPEC, for this item it was MIL-B-1953, along with a contract number, and the manufacturer's name.

Close-up of the M-1 helmet liner stamp marking. This particular helmet stamp indicates that the Micarta Division of Westinghouse Electric Company manufactured this helmet liner in 1952.

M-1 helmet liner leather chin strap, ca. 1951-1952. The chin strap holders were made of brass painted black, while the chin strap buckle was made of steel painted black.

M-1 helmet liner leather chin strap buckle, ca. 1951-1952. The chin strap buckle was made of steel painted black. Note the smooth curve design on the buckle lever.

111

BIBLIOGRAPHY

Chin Strap Buckle, 2363872, United States Patent Office, November 28, 1944

Firestone Write-up, Concerning production totals of the M-1 plastic helmet liner in World War II, Akron, Ohio, c.1951

Hawley Write-up, "Summary of the Origin and History of the Hawley Products Company", St. Charles, Illinois, August 1967

Knight, Harold A., "Fabricating and Finishing the New M-1 Soldier's Helmet", Metals and Alloys, Volume 19, Number 5, Stroudsburg, Pennsylvania, May 1944

Lewis, Frederick J., Commander, MSC, USN, et al, "Military Helmet Design", Naval Medical Field Research Laboratory, Camp Lejeune, North Carolina, June 1958

Massen, Marion, "History of the Helmet Liner", C.Q.M.D. Historical Studies: Report No. 5, Chicago Quartermaster Depot, ASF, Chicago, Illinois, April 1944

McCord Write-up, "Historical Record of McCord Corporation Contribution to World War II Ordnance Material", Detroit, Michigan, August 1945

"New Combat Helmets", Steel, Volume 111, Cleveland, Ohio, October 5, 1942

"Quartermaster Clothing and Equipment for the Tropics, Study of", Quartermaster Board Project T-356, Volume 2, The Quartermaster Board, Camp Lee, Virginia, September 1944

"Quartermaster Supply Catalog", Army Service Forces Catalog QM sec.1, OQMG Circular No. 4-Revised August 1943, Quartermaster General-ASF, Washington D.C., August 1943

"Record of Army Ordnance, Research and Development", Volume 2, Office of the Chief of Ordnance, Research and Development Service, Washington D.C., January 1946

Risch, Erna, "The Quartermaster Corps: Organization, Supply and Services", Volume 1, The United States Army in World War II, Technical Services, Office of the Chief of Military History, Washington D.C., 1953

Specifications, U.S. Army Ordnance Department, U.S. Army Quartermaster Corps, and U.S. Marine Corps 1941-1945

SPQRD 421 (Helmet, Steel M-1) 4th Ind., War Department, OQMG, Washington, D.C., February 2, 1943

SPRMD 421 (12-19-42), 00 421/17 (c) 2nd Ind., War Department, Ordnance Office, Washington, D.C., January 2, 1943

SPRMD 421.2 (18 Jan 44), (Liner, Helmet, M-1, Jungle) 2nd Indorsement, Army Services Forces, Washington, D.C., 25 January 1944

Studler, Rene R., Colonel, USA, "The New Combat Helmet", Army Ordnance, Volume 22, Washington D.C., May - June, 1942

"The Army Helmet Liner", Modern Plastics, Volume 19, May 1942

Thomson, Harry C., and, Mayo, Lida, "The Ordnance Department: Procurement and Supply", The United States Army in World War II, Technical Services, Office of the Chief of Military History, Washington D.C., 1960

"Triumph in a Helmet", Modern Plastics, Volume 20, November 1942

Wells, Daniel L., "The Story of the New American Helmet", McCord Radiator and Manufacturing Company, Detroit, Michigan, c.1946

Westinghouse Write-up, Bryant Electric, "Plastic Helmet Liners", Bridgeport, Connecticut, c.1946

Westinghouse Write-up, Micarta Division, "Helmet Liners", Trafford, Pennsylvania, c.1946

Whiting, "Statistics, United States Army in World War II", Office of the Chief of Military History, Special Staff, United States Army, Washington D.C., c.1946

"Wind-up of the Helmet", Modern Plastics, Volume 20, April 1943

SPECIFICATIONS, 1941-1945

Helmet Body
Helmet, Steel, M-1
O.D. AXS 1138
O.D. AXS 1138 Rev — Stock No. 74-H-115
O.D. AXS 1138 Rev — Stock No. 74-H-120

Helmet Liner
Lining, Helmet, Assembly for Helmet, Steel, M-1
O.D. AXS 644 — 30 October 1941

Liner, Helmet, M-1
OQMG No. 42 — 13 February 1942
C.Q.D. No. 65 — 20 June 1942
C.Q.D. No. 65a — 17 July 1942
C.Q.D. No. 65b — 02 August 1943 — Stock No. 74-L-72

Head Band
Head-Band, Assembly for Liner, Helmet, M-1
C.Q.D. No. 63 — 11 June 1942
C.Q.D. No. 63 — 20 June 1942 — amend-1
C.Q.D. No. 63a — 17 July 1942
C.Q.D. No. 63b — 06 October 1943

Band, Liner, Helmet, M-1, Head, New Type
C.Q.D. No. 63c — 10 February 1944 — Stock No. 74-B-59
C.Q.D. No. 63d — 18 August 1944

Neck Band
Neck-Band, for Liner, Helmet, M-1
C.Q.D. No. 64 — 11 June 1942
C.Q.D. No. 64a — 17 July 1942 — Stock No. 74-B-61

Neck-Band, for Liner, Helmet, M-1, Adjustable
C.Q.D. No. 64b — 15 August 1944 — Stock No. 74-B-60
U.S.A. No. 6-344 — 15 March 1945

Parachutist Helmet Body
Helmet, Steel, (Parachutist), M-2
O.D. — 23 June 1942
O.D. — 30 November 1944

Helmet, Steel, (Parachutist), M-1C
O.D. — 12 December 1944
O.D. Rev 1 — 15 January 1945
O.D. Rev 2 — 11 June 1945 — Stock No. 74-H-125

Parachutist Helmet Liner
Liner, Helmet, (Parachutist), M-1
C.Q.D. No. 66 — 17 July 1942

Accessories
Band, Helmet, Camouflage
P.Q.D. No. 255 — 02 September 1942
P.Q.D. No. 255-a — 16 October 1942
P.Q.D. No. 255-b — 18 November 1942
P.Q.D. No. 255-c — 12 April 1944 — Stock No. 74-B-56-75

Cover, Helmet, Camouflaged
Marine Corps Spec. — 17 September 1942

SPECIFICATIONS, 1950-1951

Helmet Body
Helmet, Steel, M-1
MIL-H-10990 — 05 March 1951

Helmet Liner
Liner, Helmet, M-1
MIL-L-1910 — 09 January 1950

Neck Band
Bands: Head and Neck, for Liner, Helmet, M-1
MIL-B-1953 — 31 January 1950